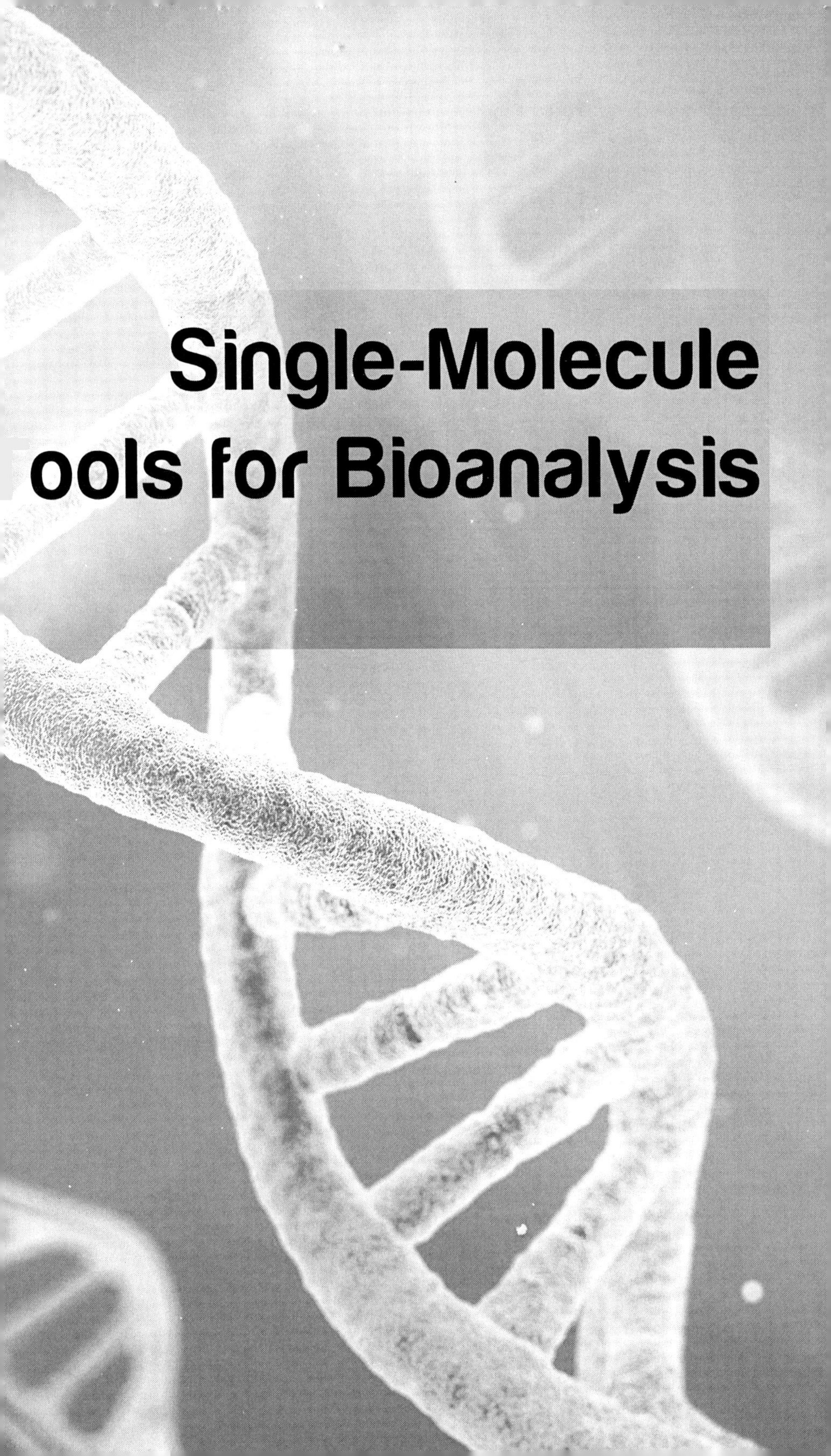

Single-Molecule
ools for Bioanalysis

Single-Molecule Tools for Bioanalysis

edited by
Shuo Huang

JENNY STANFORD
PUBLISHING

Published by

Jenny Stanford Publishing Pte. Ltd.
Level 34, Centennial Tower
3 Temasek Avenue
Singapore 039190

Email: editorial@jennystanford.com
Web: www.jennystanford.com

British Library Cataloguing-in-Publication Data
A catalogue record for this book is available from the British Library.

Single-Molecule Tools for Bioanalysis

ISBN 978-981-4800-44-0 (Hardcover)
ISBN 978-1-003-18913-8 (eBook)

Contents

Preface

Single-molecule biophysics research is a highly interdisciplinary study that requires diverse expertise in biology, chemistry, physics, and engineering, aiming to understand biological processes at single-molecule level against ensemble averaging. Investigations of single-molecule biophysics have enabled direct measurement of single-molecule properties that were not even previously feasible by any ensemble methods. These achievements include, but are not limited to, direct measurement of the elastic property of an individual strand of nucleic acids, direct manipulation of nucleic acids or protein molecules, optical imaging of cellular processes in a nanometer resolution, direct torque measurement of a supercoiled DNA, and several others. The fast development of the field has also stimulated the invention and evolution of a large variety of emerging single-molecule tools, which have enabled new concepts and applications of bioanalysis.

Though there is an urgent need to systematically summarize these achievements, it is too much for any individual review article to achieve a full coverage with sufficient details. Written by young experts in the field of single-molecule research, this book aims to provide a systematic and in-depth recap of representative topics of single-molecule bioanalysis. The book contains six chapters that cover topics on nanopores, optical tweezers, single-molecule FRET, DNA origami sensors, magnetic tweezers, and ABEL trap. Each chapter provides the general concept and a brief history of the methods, technical fundamentals, diversified forms of the methods, and the representative applications of the methods. This makes the book ideal as a textbook for a graduate-level course. In fact, the materials in this book were indeed summarized from the lecture notes of a graduate course supervised by the book editor at Nanjing University since 2015.

To make each chapter appealing to entry-level readers, a highly simplified tutorial protocol is also provided at the end of each chapter. The design of these protocols may be far from the actual measurement performed in the relevant scientific publications. They were, however, designed so that it can be finished by even undergraduate students or most graduate students with highly accessible scientific materials. Owing to the deficit of time, a few interesting topics such as electron microscopy, super-resolution microscopy, and scanning probe microscopy could not be covered. However, they are likely to be included in the future editions of this book.

Dr. Shuo Huang
School of Chemistry and Chemical Engineering
State Key Laboratory of Analytical Chemistry for Life Sciences
Chemistry and Biomedicine Innovation Center (ChemBIC)
Nanjing University

Acknowledgements

I would like to thank Dr. Mark Ian Wallace (King's College of London) for his suggestion to write this book. I would also like to thank the whole editorial team for their support during the preparation of this book. In particular, I would also like to thank Jenny Rompas and Stanford Chong, the publishers, for their support and encouragement during the book writing. I would also like to acknowledge Arvind Kanswal for the editorial support. The efforts of all chapter authors are also highly acknowledged.

Chapter 1

Single-Molecule Analysis by Biological Nanopores

Yuqin Wang and Shuo Huang
School of Chemistry and Chemical Engineering, Nanjing University, China
State Key Laboratory of Analytical Chemistry for Life Science, Nanjing, China
Chemistry and Biomedicine Innovation Center (ChemBIC), Nanjing, China
shuo.huang@nju.edu.cn

Biological nanopores are a type of proteins which form pores and were developed for in vitro single-molecule sensing. In view of their ease of use, consistency and precision of preparation, biological nanopores can be further engineered or modified for highly specialized sensing applications. Using single-channel recording, the identity of an analyte is reported from its interaction with the pore restriction during its translocation. Being geometrically compatible with single-stranded DNA or single-stranded RNA, biological nanopores, such as α-hemolysin (α-HL) or *Mycobacterium smegmatis* porin A (MspA), have long been considered the most promising candidates for third-generation sequencing. After research of ~3 decades, the prototype of a nanopore sequencer was first demonstrated in 2012 and is now widely used in a variety of genomics research programs. Sustained research of nanopore sequencing has also stimulated its

Single-Molecule Tools for Bioanalysis
Edited by Shuo Huang

ISBN 978-981-4800-44-0 (Hardcover), 978-1-003-18913-8 (eBook)
www.jennystanford.com

other applications, such as sensing of single ions, small molecules, macromolecules, biomacromolecules, or their combinations. In this chapter, we introduce the mechanism and the methodology of the biological nanopore techniques along with a tutorial protocol. We hope the reader will benefit from reading this chapter by successfully carrying out a highly simplified nanopore measurement or becoming inspired for their own research.

1.1 Introduction

1.1.1 Single-Molecule Biophysics and Nanopore

The field of single-molecule biophysics lies at the interface of physics, biology, and chemistry; its aim is to understand the mechanism of biological phenomena on a single-molecule scale. Propelled by the dramatic improvement of modern technologies such as patch-clamp technology [1], electron microscopy (EM) [2], scanning probe microscopy (SPM) [3], optical tweezers [4], magnetic tweezers [5], single-molecule fluorescence [6], super resolution microscopy [7], Förster resonance energy transfer [8], and total internal reflection fluorescence microscopy [9], researchers can now easily sense or even visualize unprecedented insights into enzyme kinetics [10], conformational dynamics [11], protein folding kinetics [12], breaking of chemical bonds [13], and ligand-binding activities [14] in a single molecule.

Each of the single-molecule methods mentioned above is irreplaceable, but not universal. For particular single-molecule applications, the proper selection of the methodology becomes critical for success. Nanopore technology is a unique method which is particularly suitable as a sensor for chain-shaped and electrically charged polymers such as nucleic acids [15]. The predecessor of the nanopore method could be traced back to the invention of the patch-clamp technology in the 1970s, which described the first instrument enabling human beings to monitor single ion channel activities on cell membranes [1]. This invention gained Neher and Sakmann the Nobel Prize in physiology or medicine in 1991 and is now widely used as a tool for electrophysiology studies of transmembrane porins or ion channels. The nanopore method, which conceptually originated from patch-clamp measurements, determines molecular identities by probing pore blockage events caused by molecular interactions of an analyte with a nanopore sensor (Figure 1.1A) [16]. In a typical nanopore measurement, a strand of a single-molecule analyte (e.g. a piece of single-stranded DNA [ssDNA]) electrophoreses through the nanoscopic aperture, generating a transient resistive pulse signal containing the molecular identity information (Figure 1.1B,C). This molecular transport process through a nanopore is termed a "translocation" event. Molecular identities are recognized by analyzing the trace fluctuations caused by the molecular translocation. The nanopore method is so sensitive that detection of subtle differences between analyte molecules is possible, making it an efficient single-molecule sensor like a miniaturized Coulter counter [17].

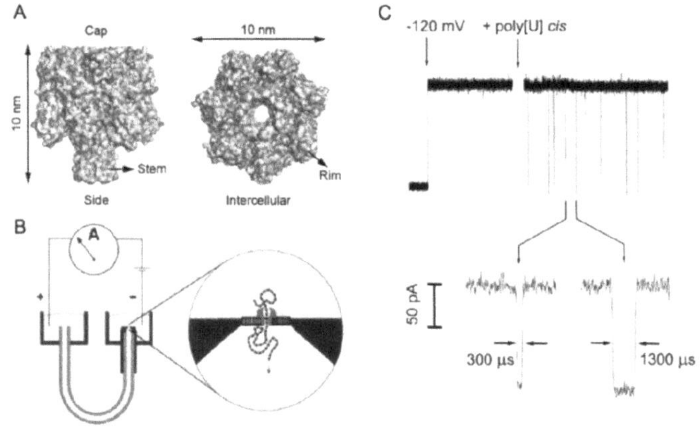

Figure 1.1 The origin of nanopore and its DNA-sensing applications. (A) The crystal structure of a heptameric α-HL nanopore. The heptameric pore appears with a mushroom shape, consisted of a wider cap (vestibule) and a narrower stem (β-barrel). The stem, which is composed of 14 antiparallel β strands, forms a cylindrical channel with a 2.6 nm diameter, permitting translocation of only ssDNA instead of dsDNA. (B) Traditional nanopore apparatus. A single α-HL nanopore can spontaneously insert into a freestanding lipid bilayer forming the only conducting path across the membrane. Analytes such as ssDNA are electrophoretically driven to pass through the pore, giving rise to the appearance of resistive pulses caused by the analyte. Reproduced with permission from ref. [18], Copyright (1999) The Biophysical Society. (C) Characteristic blockades of poly(U) translocation through an α-HL nanopore. Reproduced with permission from ref. [19], Copyright (1996) National Academy of Sciences, USA.

The nanopore measurement has advantages over other optics-based single-molecule methods because it monitors ionic current instead of photon counts. Due to limited photon emissions from fluorophores, single-molecule methods based on fluorescent microscopy such as single-molecule fluorescence resonance energy transfer (smFRET), normally produce noisy data which may limit its sensing resolution [8]. For a nanopore device based on natural ion channels, the measurement range is between 1 pA and 200 pA, which is equivalent to acquisition of 6.25–1250 million ions per second.

$$\left(1\times10^{-12}\,\mathrm{A}\right)\times\frac{1\mathrm{s}}{1.60\times10^{-19}\,\mathrm{C}}=6.25\times10^{6}\ \mathrm{ions}$$

This amount of charge transport can be reliably amplified and measured by a patch-clamp amplifier with a satisfactory signal-to-noise ratio, but prolonged excitation causes severe photo bleaching of the fluorophore, limiting the duration of the measurement. On the other hand, the nanopore device can withstand hours, or days of continuous measurement.

1.1.2 Nanopore Methods

A nanopore sensor could be generally defined as a nanoscale aperture in an impermeable membrane connecting two chambers containing electrolyte solution. A wide range of materials and methods can be utilized to make nanopore devices with different geometries and properties. An ideal nanopore sensor has to be structurally stable and geometrically consistent. To fit the cross-sectional area of a single biomacromolecule, the size of a useful nanopore sensor is normally between 1 nm and 10 nm in diameter [20]. However, nanofabrication techniques in the 1990s cannot yet reliably produce such delicate a structure over an artificial material. Until 1996, the structural determination of the *Staphylococcus aureus* α-HL by X-ray crystallography [16] suggested a biomimetic strategy to produce nanopores, and this later became an initiator of all subsequent nanopore researches.

In general, nanopore devices can be further classified into "biological nanopores" and "solid-state nanopores." All biological nanopores originate from natural transmembrane porins or their mimics, which can spontaneously penetrate a natural biomembrane or an artificial lipid bilayer and generate ion or molecular passages across the insulating membrane for biological sensing.

Biological nanopores could be massively prepared on a large scale by standard molecular biology protocols such as prokaryotic or in vitro protein expression followed with the appropriate purification steps. Though naturally composed of amino acids, biological nanopores when stored properly can stay active for a few years with no noticeable difference during measurements. It has also been experimentally verified that a biological nanopore such as an α-HL can survive an extreme of salt concentration [21], temperature [22], pH [23], and denaturants [24–26] during measurements. However, it is the fragile lipid bilayer or biomembrane which is unable to withstand harsh measurement conditions such as a high applied electrical bias, violent mechanical vibrations, or the presence of strong detergents.

Alternatively, solid-state nanopores, which are porin mimics artificially fabricated on solid-state thin materials, were developed later and aimed to provide a more durable, silicon industry-compatible solution with a complete freedom of design flexibility. Various methods such as focused ion beam [27], electron beam [28], track-guided chemical etching [29], and dielectric breakdown [30] could be used for pore drilling. Solid-state nanopore techniques offer advantages of a more flexible pore geometry and patterning along with a variety of surface property modifications but suffers from a poor biosensing performance due to the inconsistency of pore manufacturing at the nanometer scale.

Other emerging nanopore technologies, which cannot be classified in either of the types mentioned above, have also been investigated. For instance, the fusion of biological nanopores and solid-state nanopores, termed hybrid nanopores, is expected to overcome the limitations on both sides [31, 32], and the emerging DNA origami nanopores possess the ability for precise control over the geometry and the surface functionality [33–35].

Among different types of nanopores, the biological nanopore method is the first reported method in the field and is currently the only nanopore method which can sequence DNA or RNA. With long-term debates over the pros and cons of different pore types, biological nanopores currently still outperform their solid-state counterparts in the aspects of the signal-to-noise ratio, spatial resolution, and manufacturing consistency. Due to the limitations of space, the remainder of the discussion will be focused on biological nanopores.

1.1.3 Biological Nanopores

Biological nanopores are a category of transmembrane porins with a considerably large pore lumen measuring from 1 to 5 nm in diameter, which are utilized for single-molecule sensing. Reported biological nanopores including *S. aureus* α-HL [19], MspA [36], *Escherichia coli* ferric hydroxamate uptake protein A (FhuA) [37], bacteriophage phi29 connector [38], *E. coli* cytolysin A (ClyA) [39], *E. coli* outer membrane protein G (OmpG) [40], *E. coli* outer membrane protein F (OmpF) [41], *Actinia fragacea* fragaceatoxin C (Frac) [42], *E. coli* curli production assembly/transport component CsgG (CsgG) [43], and human specificity protein 1 (Sp1) [44] could be similarly utilized as single-molecule sensors (Figure 1.2). The α-HL, which is the most studied biological nanopore, is the most robust nanopore sensor used to date. In nature, it is an exotoxin secreted by the human pathogen *S. aureus* bacterium. In its heptameric form, α-HL appears as a mushroom-shaped protein (with a cap domain and a stem domain) and a molecular weight of 232.4 kDa. The stem domain, which is embedded in the lipid membrane, is composed of 14 antiparallel *β* strands that form a cylindrical channel for molecular transportation. The narrowest spot of the cylindrical channel is ~1.4 nm in diameter. It serves as the recognition site for molecular identity discrimination [16] and permits passage only of ssDNA. The cap domain, which has an inner diameter of ~4.5 nm, is capable of accommodating a short fragment of dsDNA.

The MspA nanopore, which is a funnel-shaped octameric pore with a molecular weight of 157 kDa, is more useful in nanopore sequencing. Benefiting from its short and narrow recognition site, which is ~0.6 nm long and ~1.2 nm wide, the MspA pore is capable of reading a frame of only 4 nucleotides at a time over a piece of ssDNA during its translocation.

Both α-HL and MspA permit translocation of ssDNA but not dsDNA. The phi29 connector protein, which is a dodecamer of the GP10 protein, opens up a ~3.6 nm diameter channel capable of translocating dsDNA or proteins of a low molecular weight. However, it is reported that unlike α-HL and MspA, a phi-29 connector protein cannot spontaneously insert into the lipid bilayer but requires assistance from vesicle fusion [38]. However, vesicle fusion may lead to multichannel insertion or an unknown channel insertion orientation.

As transmembrane proteins with known sequence and structure, the geometry or charged residues of biological nanopores could be slightly but precisely modulated by site-directed mutagenesis [45] for particular biosensing applications [46]. However, site-directed mutagenesis is not always successful and normally leads to significant efforts in the screening of desired mutants.

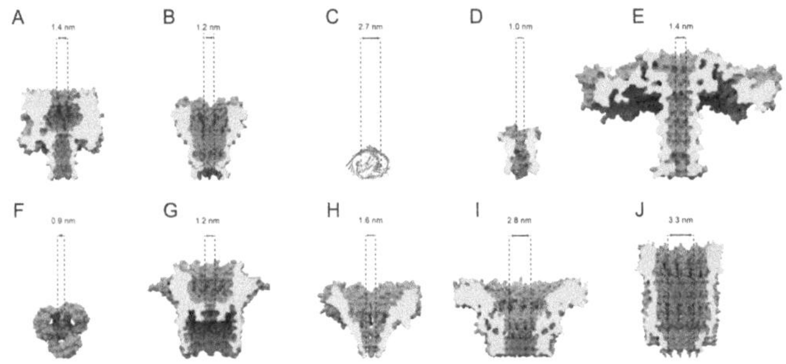

Figure 1.2 Crystal structures of different biological nanopores. (A) α-HL (PDB: 7AHL) [16]. (B) MspA (PDB: 1UUN) [47]. (C) FhuA (PDB: 1BY3) [48]. (D) OmpG (PDB: 2F1C) [49]. (E) Aerolysin (PDB: 5JZT) [50]. (F) OmpF (PDB: 2OMF) [51]. (G) CsgG (PDB: 4Q79) [43]. (H) FraC (PDB: 4TSY) [42]. (I) phi 29 connector (PDB: 1FOU) [52]. (J) ClyA (PDB: 6MRT) [53].

1.2 Methodology

1.2.1 Preparation and Engineering of Biological Nanopores

A key advantage of a biological nanopore over its solid-state counterpart is that it is extremely easy to prepare in a large quantity but with a low cost. The chemical nature of a biological nanopore can be manipulated with atomic precision and consistency. As has been reported, biological nanopores can be made in vitro using standard protein expression protocols [46] or in a prokaryotic expression system [36].

Taking α-HL as an example, the plasmid DNA containing the target gene encoding the protein monomer is custom designed and synthesized by commercial services (Genescript, New Jersey). This plasmid DNA could be further copied to develop the quantity and purity needed by a routine plasmid mini, midi, or maxi prep kit (Qiagen, QIAprep Spin Miniprep Kit).

A quick method of protein expression is by in vitro transcription and translation (IVTT). Following the standard protocol provided by the IVTT kit (TnT Quick Coupled Transcription/Translation System, Promega), ~500 ng of protein monomer can be biosynthesized in 2 h. The prepared protein monomer is then incubated with rabbit red blood cell membrane fragments at 37 °C for a further 2 h to assist protein oligomerization. The oligomerized protein sample is then purified and recycled by polyacrylamide gel electrophoresis and radioactive gel exposure. The whole process takes around 1.5–2 days to completion which is convenient for protein-screening purposes. However, the IVTT method has disadvantages of a low yield and the requirement of radioactive labeling during purification.

An alternative means of preparing biological nanopores is by prokaryotic expression followed with purification by fast protein liquid chromatography. In this approach, a tag protein such as His-tag [54] or Strep-tag [36] is normally placed on either terminus of the target protein for later purification purposes. The prokaryotic protein expression system can efficiently produce several mg of protein in 3–4 days. However, if the target protein is cytotoxic to the host cell during expression, difficulties may ensue. This method also requires more complicated purification procedures to eliminate interferences from the background proteins generated by the host cell. Site-directed mutagenesis is sometimes needed to optimize the performance of the protein nanopore [55]. This can be accomplished by direct synthesis of the mutated gene or with a site-directed mutagenesis kit.

The prepared protein nanopore is dissolved in Tris-EDTA buffer at pH 7.0 along with detergent to avoid protein precipitation. If properly stored at -80°C, the biological nanopores can stay active for up to 10 years and it has been suggested that protein nanopore samples should divided into small aliquots (1 μL each in PCR tubes) for long-term storage so that freezing and thawing of the sample is minimized.

1.2.2 The Instrument and the Device

A complete nanopore measurement platform includes a data acquisition module, a noise-insulating module, and a measurement module. For the data acquisition, most laboratories rely on commercial patch-clamp amplifiers (Axopatch 200B, Heka EPC 800, Elements Eone, or Chimera VC100) along with a compatible digitizer for analog/digital conversion and a computer for instrument coordination and trace recording. Some laboratories prefer custom-made amplifiers for high bandwidth measurement applications or ease of device integration [56].

In a nanopore measurement, the ionic current through a single nanopore is approximately 0–200 pA for biological nanopores or 0–10 nA for solid-state nanopores. The patch-clamp amplifier serves to convert analog current signal into analog voltage signal, which is then acquired by the digitizer in a binary format. The digitizer, which is either integrated with the patch-clamp amplifier (EPC 800, Heka) or provided as a separate component (Digidata 1550A1, Molecular Devices), may also be replaced with a data acquisition card (National Instrument USB-6210) or even a digital oscilloscope for cost saving or expandability purposes.

In our laboratory, we use an Axopatch 200B patch-clamp amplifier paired with a 1550A1 16-bit digitizer (Molecular Devices) with a dynamic input range of ±10 V. The 16-bit digitizer serves to divide the signal with 2^{16}-1 (65535) digitized bits, generating a resolution of 0.305 mV/bit ($20V/2^{16}$). The gain of the amplifier could be set higher to further improve the measurement resolution but at the expense of the dynamic range or vice versa. For biological nanopore measurements, the gain is set to 10× or higher, which provides a dynamic range of ±1 nA and an improved resolution of 0.0305 pA/bit, which is ideal for discriminating between minimally different molecular analytes.

The noise-insulating module is critical for isolation of external electrical and mechanical noises from a sensitive nanopore experiment. The radiative electrical noise can be effectively shielded if the measurement is carried out in a conducting enclosure (a shielding box or Faraday cage) electrically connected to the common ground of the patch-clamp amplifier. Mechanical vibrations from external sources such as music, walking, talking, and keyboard typing can also be transmitted into the recorded signal and appearing as low frequency fluctuations of the signal baseline. For optimum performance, in our laboratory, the Faraday cage is bolted on an optical table with air floating supports, which effectively isolates it from most mechanical vibrations above 2 Hz.

A minimal measurement module is composed of a pair of Ag/AgCl electrodes, a measurement chamber and other accessories. Pure silver wire 1.5–2 mm in diameter, 2 cm in length is first soldered to an electrical jump cable (maximum 10 cm) at one end, then surface polished to eliminate the oxidized layer before being immersed in NaOCl solution (1–5% (w/v) solution in water) for 3–5 h until a thick, dark gray colored layer forms over the silver surface. The other end of the jump cable is soldered to a male connector pin for an electrical connection to the head stage of the patch-clamp amplifier. The measurement chamber is normally made of electrical insulating Teflon or polyformaldehyde polymers with 2 chambers (*Cis* and *Trans*) of 50–1000 μL on each side to accommodate the electrolyte solution, for example, 1 M KCl. The two chambers are geometrically connected with an aperture ~1 cm in diameter and hold a chip containing a solid-state nanopore or a Teflon film for lipid bilayer formation in biological nanopore experiments. During the measurement, the electrically connected Ag/AgCl electrodes are separately immersed in the electrolyte solution on each side of the chamber to form a closed circuit and the data acquisition for a nanopore measurement can then commence.

There are also other accessories such as peltier stages (temperature controls), magnetic stirrers (analyte mixing), or syringe pumps (automated liquid exchange) to facilitate the measurement. However, turning off these instruments during the recording is suggested as they could introduce noise that can also be picked up by the preamplifier.

1.2.3 The Electrochemistry Mechanism

For any type of nanopore, two chambers filled with electrolyte solution are separated by an impermeable membrane containing a single nanopore, which is the only pathway by which ions and molecular analytes can be transported across the membrane. A pair of Ag/AgCl electrodes electrically extended from the patch-clamp amplifier are placed on each side of the chamber, in contact with the electrolyte solution in either side. A potential bias is applied from the patch-clamp amplifier to promote directional flowing of ionic species through the pore. The corresponding electrochemical reaction on the electrode surface is: $Ag(s)+Cl^- \leftrightarrow AgCl(s)+e$.

The gain of an electron on the cathode electrode results in the release of a Cl^- ion to the solution. On the other side, the gain of a Cl^- on the anode electrode results in the release of an electron to the electrical circuit and thus generates a sustainable flow of ion across the membrane. Conventionally, the chamber which is electrically grounded is defined as the *Cis* side, while the other one which is electrically biased is defined as the *Trans* side.

Although the pair of Ag/AgCl electrodes are relatively far (>2 cm) from each other, the potential drop between the electrodes is mostly in the vicinity and on the scale of the thickness of the membrane of the nanopore due to the large resistance (10^8 ohm/cm^2) of the membrane, the lipid bilayer for biological pores or synthetic membrane for solid-state pores. Thus, the electric field, which is opposed to the gradient of the electrical potential ($E = -\nabla\varphi$), becomes stronger as it moves closer to the nanopore restriction. The electrophoretic force acting on the analyte is proportional to the electric field ($F = Eq$) and is strong enough against the Brownian motion of the molecule only within a hemispherical area in front of the nanopore. Thus the analyte molecule, for example, ssDNA, has to achieve a high enough final concentration (>200 nM) so that a reasonable amount of the event may be observed.

1.2.4 The Nanopore Measurement

In a typical nanopore measurement, a voltage is applied across the membrane by a pair of Ag/AgCl electrodes. The two chambers filled with electrolyte solutions (i.e., 1 M KCl, 10 mM HEPES, pH 7.0) are separated by a ~30 μm thick Teflon film. An aperture ~100 mm in diameter, coated with hexadecane, is located in the center of the film and serves as a solid support for the lipid bilayer assembly. The lipid bilayer can be formed by pipetting the solution in either chamber "up and down" by forming Langmuir–Blodgett lipid monolayers on each side of the aperture. A triangular wave with a slope of 1 V/s is often applied to differentiate between a successfully formed bilayer and a permanent blockage caused by air bubbles or contaminants. Free nanopores in solution can spontaneously insert into the lipid bilayer forming the only electrical connection between the two electrolyte solutions. After the first pore insertion, the chamber should be exchanged, or perfused with fresh, nanopore free buffer to "wash away" excess biological nanopores and so avoid further pore insertions. The buffer exchange process should be accomplished with care and patience since abrupt mechanical vibration could potentially cause bilayer rupture and rapid exchange of the electrolyte solution in the chamber could sometimes assist further pore insertions. A voltage applied across the nanopore causes ions to flow through the nanopore, establishing a measurable ion current. The whole process is detailed in the tutorial protocol, below.

1.2.5 Measurement Noise and Bandwidth

In the time domain, the acquired ion flow through an open nanopore generates a root-mean squared noise centered about a steady DC signal. The power spectrum density analysis shows a 1/f noise in the lower frequency regime (below ~300 Hz or depending on the pore materials) and a positive power dependence on frequency in the higher frequency regime (above ~300 Hz or depending on the pore materials) [57]. This increased noise contribution from the high frequency end limits the bandwidth of the nanopore measurement, and this means that at some point in the frequency domain, the noise contribution will be too significant for the signal to appear in the time trace. The main source of the noise in the high-frequency region is the membrane capacitance of the nanopore device. This can be optimized by reducing the membrane capacitance of a solid-state nanopore [58] or minimizing the lipid bilayer area of the biological nanopore. More effectively, the high-frequency noise can be eliminated by a low-pass Bessel filter if the event of interest is significantly slower than the cut-off frequency of the filter. For this reason, much effort has been paid by researchers to slow down the speed of analyte translocation in order to resolve the signals of interest.

Normally, the cut-off frequency should be greater than five times the inverse of the mean event time [59]. According to the Nyquist–Shannon sampling theorem, the sampling rate should be more than twice the frequency of interest. In practice, it is suggested to over sample 10 times to preserve the signal shape.

1.2.6 Data Analysis

Digitally acquired time traces are normally produced in an .abf binary format. During a typical DNA translocation experiment, there could be more than thousands of translocation events to be analyzed for a single experiment condition, and it is necessary to perform the data analysis by computer. The Clampfit software accompanying the Axon 200B patch-clamp system is a general purpose data analysis program with some integrated statistical analysis functions. Routine data analysis such as I–V curve analysis, step recognition, and analysis of all-points histograms can be performed directly in Clampfit.

For single-channel traces containing pore blockage events, the key parameters to be extracted are the residual current I_b, the dwell time t_d, and the interevent interval t_{on}. An automatic search can be done with the "single-channel search" function of Clampfit and when the results are exported to professional plotting software such as Origin or Igor, they can be eventually plotted as event histograms. The event histogram of I_b normally fits well with a Gaussian function and t_d fits a Gaussian function with an exponential tail.

However, Clampfit is not always sufficient for complicated event extractions. Data from some nanopore blockage events such as sequencing [60], structural unzipping [61], or unfolding data [62] appear as a continuous step-like signal which can be difficult for Clampfit to distinguish. In these cases, writing a custom program (e.g., with Matlab, Python, R, or LabView) according to the characteristic shape of the events of interest for automated event extraction is recommended. There are also many useful online resources on www.thenanoporesite.com/nanopore-software.html available for free download.

1.3 Applications

1.3.1 DNA Sensing and Sequencing

1.3.1.1 Free translocation of polynucleotide molecules

In 1996, Kasianowicz et al. first demonstrated the phenomenon of single-molecule transport of polyuridine oligomers through an α-HL nanopore with a continuously applied potential [19]. Subsequently, the crystal structure of the α-HL nanopore was reported. ssDNA translocation through the pore in a strictly single-file and sequential order was clearly indicated on a structural basis [16], and Kasianowicz and a few others proposed the concept of nanopore sequencing, from which the sequence of current modulations induced by the DNA translocation could be aligned with its base composition. However, DNA sequencing still remains a challenge due to a lack of spatiotemporal resolution from the pore and the acquisition system.

First, the translocation speed of DNA, at 1–22 μs/nt [18, 63], is always too fast, and individual bases are hardly resolvable due to the limited bandwidth of existing patch-clamp amplifiers. Second, a single-nucleotide resolution cannot be achieved as the current blockades are found to be the consequence of ~10–15 nucleotides that are simultaneously accommodated within the long β-barrel of an α-HL nanopore [20]. Therefore, it has proved to be impossible to directly sequence DNA simply by analyzing translocation events acquired with α-HL nanopores.

1.3.1.2 Discrimination of immobilized nucleic acids in the biological nanopores

To enable prolonged measurements, ssDNA can be directionally immobilized and electrophoretically stretched within a nanopore by forming either a terminal hairpin [64] or a biotin-streptavidin complex [65]. Using this strategy, clear discrimination of single base substitutions of A, C, G in a homopolymeric strand of thymidine, was demonstrated using an α-HL nanopore [65]. Notably, the common epigenetic modifications in the DNA strand, such as 5-methylcytosines and 5-hydroxymethylcytosines, can also be distinguished [66], as can epigenetic ribobases [67]. However, the recognition sites of the α-HL nanopore are still insufficiently sharp to distinguish all bases in a diverse context.

MspA, which was pioneered by Gundlach et al., possesses a short constriction 1.2 nm in width and 0.6 nm in length, indicating an ideal geometry as a high-resolution DNA reader for nanopore sequencing. By performing static measurement with DNA homopolymers with a tethered streptavidin stopper [68], Gundlach et al. observed a signal discrimination with an order of magnitude larger than those found with the α-HL nanopore, which confirmed that MspA is superior to α-HL as the nanopore sensor for sequencing.

1.3.1.3 DNA strand sequencing

Many attempts were made to slow down the speed of DNA translocation, including control over the temperature [63, 69], manipulation of the voltage [70], modification of analytes [71], and many others [15, 72]. Though one to two orders of magnitude decrement in the translocation speed were achieved, this was at the cost of either a largely suppressed channel current or a much reduced signal-to-noise ratio [15, 69, 73]. A desired DNA translocation modulation was recently enabled by coupling the DNA template with an enzyme motor adjacent to the pore orifice. With this strategy, a highly processive enzyme can ratchet along the DNA template one nucleotide at a time with a spacing of milliseconds and can replicate up to tens of thousands of nucleotides, promising a single-nucleotide pace of reading and a long read length during extended measurements.

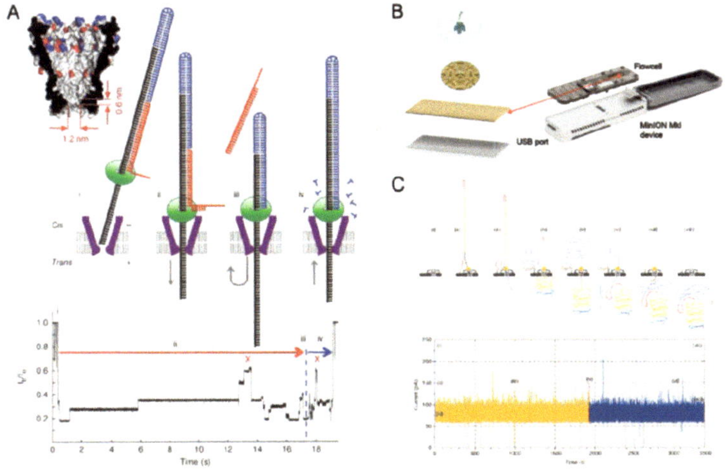

Figure 1.3 Nanopore sequencing. (A) A prototype of nanopore sequencing using an MspA nanopore. A phi29 DNAP, which serves as a motor protein, was employed to control the forward and reverse ratcheting of DNA templates through MspA. Sequence-dependent current levels were observed when the DNA template was translocating through the pore. Reproduced with permission from ref. [60], Copyright (2012) Nature Publishing Group. (B) The Minion nanopore sequencer. Each consumable flow cell generates as much as 512 nanopores for parallel DNA sequencing. Reproduced with permission from ref. [74], Copyright (1969) Elsevier. (C) Sequencing data reported from the MinION nanopore sequencer. Reproduced with permission from ref. [75], Copyright (2016) The Author(s).

Among the selection of the motor enzymes, the bacteriophage phi29 polymerase (phi29 DNAP), as one of the B-family DNAPs, has been shown to be superior for its remarkable processivity, which is ≥70 kilobases in vitro and three to four orders of magnitude affinity for DNA substrates more than that of a Klenow fragment [76]. In this system, the DNA strand translocates through the α-HL nanopore in a two-step manner: voltage-driven unzipping, wherein the DNA strand electrically threads through the pore in 5′–3′ direction at a velocity of 2.5 nucleotides per second, and replication-driven ratcheting, wherein the deprotected DNA strand was pulled through the pore in 3′–5′ direction at a velocity of 40 nucleotides per second. In this way, individual DNA template can be controlled to forward and reverse ratcheting through the pore. However, the registry error rate of any single position is up to 10–24.5%, resulting in significant base insertions or deletions relative to the correct sequence [77].

Gundlach et al. conducted the same measurement using a MspA nanopore (Figure 1.3A). [60]. A two-step current pattern was observed in the opposite time order, but each distinct current level could be better resolved with millisecond accuracy (~28 ms) and pico-ampere (pA) precision, and could be clearly aligned to a known DNA sequence [60]. Following this strategy, the DNA sequence from the bacteriophage phi X 174 genome relying on nanopore readouts was reported [78]. This approach has also been employed for detection of single-nucleotide polymorphisms [78] and epigenetic modifications [79].

Although the phi29 DNAP-mediated MspA nanopore sequencing integrates the DNA speed control and single base resolution admirably, the sequencing error rate cannot be eliminated owing to the backsteps and stochastic motion of the phi29 DNAP. Studies on enzyme motors with sharply improved performance, such as a helicase that advances the DNA template strand captured by the MspA nanopore with subangstrom resolution, have been accomplished [80].

In 2012, Oxford Nanopore Technologies released the first nanopore-based DNA sequencer, named MinION (Figure 1.3B,C) [81]. As advertised, this disposable sequencing gadget, which costs <$1000, is capable of exporting a gigabase of DNA sequences per day [74, 75, 82, 83]. However, the poor accuracy of the base identification remains a major drawback, limiting its broad implementation over existing sequencing technologies. In 2016, based on several measurements of replicated circular DNAs, the latest MinION successfully sequenced the Ebola virus genomes stemming directly from patients' samples [84]. Acknowledging a portable size and a constantly improved accuracy, a nanopore-based sequencer is anticipated to be widely applied as a diagnosis tool in clinical trials.

1.3.1.4 Exosequencing

Exosequencing has been demonstrated as a parallel alternative strategy for nanopore sequencing with a high-base calling accuracy. In this approach, an exonuclease is conjugated adjacent to the nanopore rim, and digests individual nucleotides from a DNA template to be sequenced. These nucleotides, when passing through the nanopore in sequence according to their cleavage order, can be sequentially identified [85–87]. Studies toward an exosequencing system show that four nucleoside monophosphates can be discriminated with an average accuracy up to 99.8% [87], as can the four ribonucleoside diphosphates [88]. However, it is hard to guarantee the order of the released nucleotides against the interference from stochastic diffusion.

Another alternative exosequencing platform is Genia's NanoTag sequencing technology. In this approach, four bases can be distinguished indirectly by the passage of four differently sized ethylene glycol (PEG)-coumarin tags released from the bases as they incorporate into a DNA template. This has overcome the inherent limitation in the spatial resolution of a biological nanopore [89, 90]. A low-cost Genia sequencer claims, as advertised, that sequencing a complete human genome could cost less than $100.

1.3.2 Efforts toward Protein Sequencing

1.3.2.1 Translocation of peptides

The success of nucleic acid sequencing by nanopore has led to the possibility of nanopore sequencing of other biomacromolecules, such as peptides, proteins, or polysaccharides [91]. Though the 20 amino acids have significant differences in their physiochemical properties, direct protein sequencing using nanopore still faces challenges from a large entropic barrier associated with unfolding a protein before it translocates through the pore. In contrast to nucleic acids, which are homogeneously charged on their phosphate backbone, an electrically neutral biomacromolecule, such as a protein, is in principle difficult to be electrophoretically trapped and stretched inside a nanopore.

Preliminary attempts have been made however in nanopore sensing of short polypeptides [92–95], and synthetic peptides with different structures, such as single, double, and collagen-like triple helices, were easily distinguished in the characteristic nanopore readouts (Figure 1.4A) [92]. An interaction between positively charged signal peptides and a β-barrel pore was also observed, which could be instructive for the modulation of protein capture and translocation [93].

1.3.2.2 Translocation of unfolded proteins

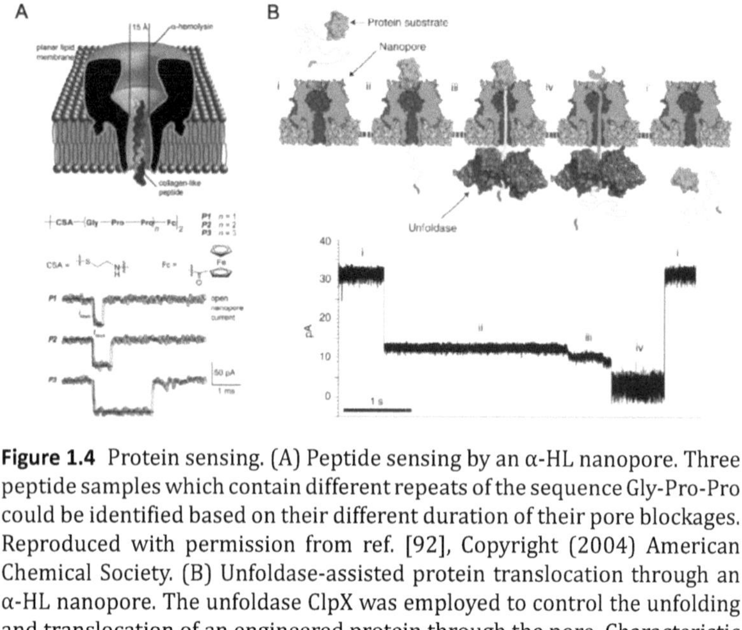

Figure 1.4 Protein sensing. (A) Peptide sensing by an α-HL nanopore. Three peptide samples which contain different repeats of the sequence Gly-Pro-Pro could be identified based on their different duration of their pore blockages. Reproduced with permission from ref. [92], Copyright (2004) American Chemical Society. (B) Unfoldase-assisted protein translocation through an α-HL nanopore. The unfoldase ClpX was employed to control the unfolding and translocation of an engineered protein through the pore. Characteristic events of protein unfolding could be observed. Reproduced with permission from ref. [62], Copyright (2013) Nature Publishing Group.

Strategies to unfold proteins prior to translocation have also been examined by performing measurements in the presence of denaturants, such as urea [25] and guanidinium chloride (GdnHCl) [24, 26, 96]. With the assistance of denaturant, translocations of unfolded proteins were observed when urea and GdnHCl were applied with 4 M and 1.5 M final concentrations, respectively. Fortunately, the protein nanopore used in these assays was unaffected by the presence of such highly concentrated denaturants. However, it is the fragile lipid membrane which cannot withstand strong denaturants like sodium dodecyl sulfate (SDS) [97, 98]. Measurements at a high temperature have also been used to unfold proteins during a nanopore assay. However, in this case, the translocation speed would be further increased thermodynamically, and clear discrimination of any sequence-dependent information during the assay would not be possible [99]. Alternatively, single-protein molecules could be mechanically forced to unfold with the assistance of the electrophoretic force. Linear poly-anionic tags, such as oligonucleotides, could be used to tether the C or N terminus of a protein to facilitate the threading of the terminal end into the nanopore [100]. However, the reported nanopore readout still fails to report clear sequence dependence as protein translocation is happening in a diffusion limited manner.

1.3.2.3 Enzyme-assisted unfolding during nanopore translocation

Analogous to nanopore-based DNA sequencing, an enzyme-assisted strategy was demonstrated by Akeson et al. to achieve processive unfolding of a protein during translocation through the α-HL nanopore (Figure 1.4B) [62]. Fueled by ATP hydrolysis, the unfoldase Caseinolytic protease X(ClpX) can specifically bind to the unique polyanion tags of a protein, and can subsequently ratchet along the protein at the speed of 80 amino acids per second. Segments of a single protein could be discerned based on sequence-dependent features when the protein slides through the α-HL nanopore. This proof-of-concept study demonstrates a promising way to achieve protein sequencing. However, follow-up reports using this strategy have rarely appeared, which may indicate that protein sequencing using nanopores still faces other unsolved technical issues.

1.3.3 Sensing of Small Molecules and Single-Molecule Chemistry Intermediates

Besides sequencing nucleic acids or possibly a variety of other biomacromolecules, nanopore has performed well in sensing of single smaller sized molecules. In general, direct translocation of small-sized molecular analytes through a biological nanopore results in almost no detectable events due mainly to the fast speed of translocation. However, when chemical binding between a small analyte and its receptor is established in the lumen of an engineered nanopore, fluctuations in the current may be observed. On this basis, a variety of engineered biological nanopores have been constructed as stochastic sensors for the identification and quantitation of small analytes, such as organic molecules [86, 101–108], cations [109], and metal ions [110–113].

There are two strategies for the detection of small analytes by nanopore stochastic sensing. First, a molecule adapter specific to the analyte may be engineered within the lumen of the pore through noncovalent interaction or site-directed mutagenesis. β-cyclodextrin (βCD) and its derivatives have been used widely in this scenario, which can lodge in the lumen of the α-HL pore reversibly and provide a hydrophobic cavity suitable for a variety of guest molecules, such as adamantine derivatives [102], enantiomers [104], mustards [86], and nerve agents [105]. For continuous detection, covalent attachment between β-cyclodextrin and the pore lumen was performed with α-HL mutants [86]. On the other hand, pore restrictions established by site-directed mutagenesis were engineered to interact directly with divalent metal ions including Zn(II), Co(II), and Cd(II) [110]. Various nitroaromatics such as 2,4,6-trinitrotoluene were similarly sensed [103].

The second strategy is based on the observation of a prolonged duration induced by the binding event of an analyte and its DNA aptamer. When bound with an analyte, the specific DNA aptamer can be transformed spontaneously into a stable duplex, such as three-way junction, hairpin, or a g-quadruplex, which directly reports a clear discrimination in subsequent nanopore readouts. This approach allows for the highly selective and sensitive detection for trace amounts of cocaine [101], mercury (II) [111, 112], lead (II) [113], and barium (II) [113].

Engineered nanopores have also been used as nanoreactors for the observation of chemical binding events at a single-molecule level (Figure 1.5A) [114, 115]. By monitoring the telegraph fluctuation in the ionic current through an engineered α-HL pore, events resulting from single bond-formation or breaking were observed, and reveal rich information about the stoichiometry, kinetics, and intermediates of the chemistry that cannot be observed by other means [116]. With this concept in mind, the coordination reactions between organoarsenic (III) compounds with the thiol of cysteine [117–119] as well as Au(I) compounds with the imidazole of histidine [120] were observed. Two approaches have been employed to extend single-molecule chemistry studies beyond those involving natural amino acid residues. One method is to modify the covalently connected active groups on the inner surface of nanopores with cysteine, which has been used to monitor the step-by-step growth of a polymer chain [121], the photoisomerization of an azobenzene [122] or the covalent interactions of boronic acids with diols [123]. In another approach, the active groups were introduced by incorporating a fragment of an unnatural amino acid by native chemical ligation. This allows introduction of an alkyne side chain [124] or a ketone side chain [125] into synthetic α-HL nanopores for the observation of the click reaction and oxime chemistry.

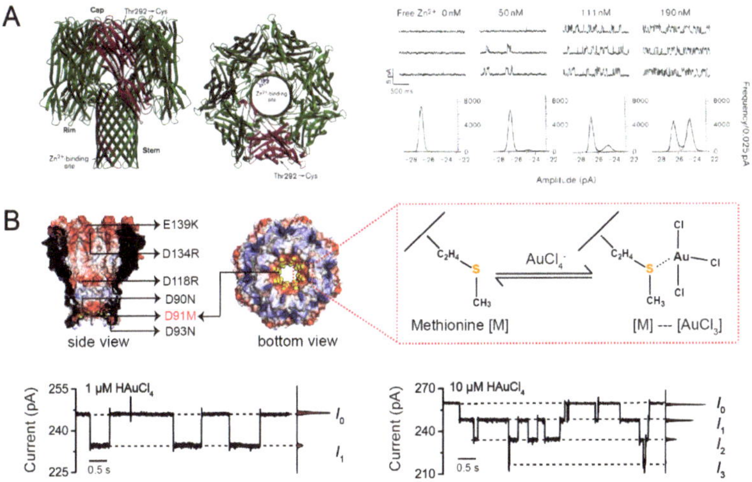

Figure 1.5 Single-molecule chemistry observed in a nanopore reactor. (A) The structure of an engineered α-HL nanopore and its coordination interaction with Zn^{2+}. A Zn^{2+} binding site was introduced into the lumen of the α-HL nanopore by site-directed mutagenesis. Reversible bindings of Zn^{2+} with the pore restriction report corresponding event signatures. Reproduced with permission from ref. [114], Copyright (1997) Elsevier. (B) The structure of an engineered MspA nanopore and its coordination interaction with $[AuCl4]^-$. The mutant MspA introduced with eight identical methionine residues at position 91 could bind one or multiple $[AuCl4]^-$ ions simultaneously. The conical geometry of MspA reports much enlarged single-molecule chemistry events. Reproduced with permission from Ref. [115], Copyright (2019) Nature Publishing Group.

However, all reported single-molecule chemistry studies were carried out by engineered α-HL mutants, which suffer from an extremely poor event amplitude and a blunt pore geometry. Recent reports of MspA mutants with a designed reactive site suggest that a conical geometry may be superior when reporting giant single-molecule chemistry events with, however, a much reduced requirement of pore engineering (Figure 1.5B) [115].

1.3.4 Single-Molecule Enzymology

Single-molecule enzymology is a branch of biochemistry, which deals with the properties, the functions, and the mechanism of naturally occurring enzymes down to a single-molecule scale. Emerging techniques such as optical tweezers [126, 127], magnetic tweezers [128], atomic force spectroscopy [129], and smFRET [8] have been widely applied in this field. However, the aforementioned techniques still lack sufficient spatiotemporal resolution to fully resolve conformational fluctuations on the submolecular scale, which is critical for an in-depth understanding of the functional mechanism of enzymes.

Biological nanopores, which are nanocavities with well-defined structural and physiochemical properties, can be engineered to monitor single-molecule enzymatic activities with a high spatiotemporal resolution. The only remaining task is how to design a nanopore assay so that enzymatic activities could be reported as detectable ionic current fluctuations. Inspired by attempts to achieve nanopore sequencing for a variety of biomacromolecules, motor enzymes that move processively along DNA, RNA, or protein substrates [130, 131] have become the most suitable analyte in nanopore single-molecule enzymology studies. When properly designed, single-molecule kinetics of enzymatic binding [132], DNA replication by a Klenow fragment [133], a T7 DNA polymerase [134] or a phi29 DNA polymerase [76], enzymatic cleavage by a Exonuclease I [87], or ClpX protease [62] could be studied with α-HL nanopores. However, the mushroom-shaped α-HL nanopore still lacks a spatial resolution superior to other existing single-molecule enzymology methods.

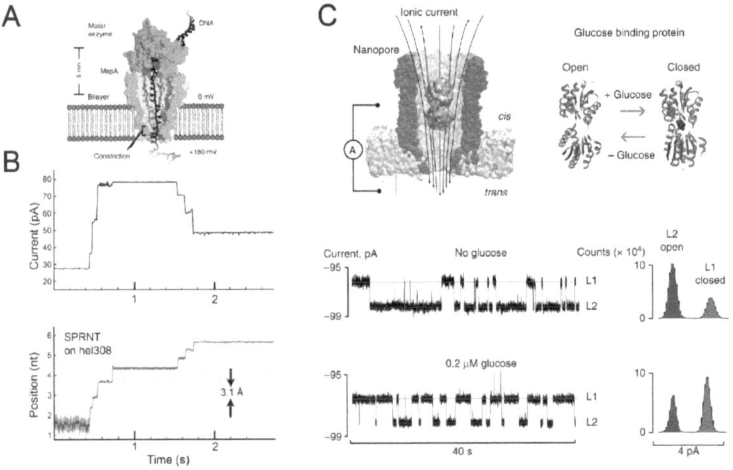

Figure 1.6 Nanopore enzymology. (A) The schematics of single-molecule picometer-resolution nanopore tweezers (SPRNT). Measurements of enzymatic location with a picometer resolution were enabled by measurements with a piece of ssDNA with a known sequence. However, SPRNT is currently limited to measurements with only nucleic acid binding enzyme, with which a strong binding affinity is normally preferred. Reproduced with permission from ref. [80], Copyright (2015), Nature Publishing Group. (B) Conversion of current trace to DNA displacement during SPRNT. (Top) Current levels versus time trace for a single Hel308 molecule moving on a DNA template. (Bottom) Corresponding enzyme position versus time. Reproduced with permission from ref. [135], Copyright (2016) Elsevier. (C) A nanopore glucose sensor. The structure switch of a glucose-binding protein could be dynamically monitored when the enzyme was lodged inside a ClyA nanopore. The concentration of glucose in a body liquid sample could be subsequently measured. Reproduced with permission from ref. [136], Copyright (2018) Nature Publishing Group.

An emerging technology named single-molecule picometer-resolution nanopore tweezers (SPRNT) [137] has been recently developed for real-time observation of enzyme kinetics. The core component of SPRNT is a geometrically sharp biological nanopore MspA, which had been previously developed for nanopore sequencing [60]. In SPRNT, a strand of synthetic DNA with a known sequence serves as a molecular ruler. When bound with a motor enzyme, such as a helicase or a DNA polymerase, the DNA is electrophoretically driven into an engineered MspA nanopore, activating the progression of the DNA with a translocation speed of 1–100 nt/s through the enzyme (Figure 1.6A) [60]. Owning to the high spatial resolution of MspA, the measured ionic current readout directly reveals the relative position of the enzyme in reference to the DNA ruler, and a ~40 pm and a submillisecond spatiotemporal resolution could be achieved simultaneously (Figure 1.6B) [80, 138]. However, with a requirement of a high processivity from the enzyme, SPRNT is in principle limited to DNA/RNA binding enzymes with a high binding affinity to their nucleic acid substrates.

A general and more straightforward approach to performance of nanopore enzymology is to monitor dynamic structural transition of an enzyme from direct nanopore readouts. This has been performed either by engineering a catalytic protein nanopore [139] or by monitoring the translocation kinetics of enzyme molecules before or after ligand binding [140]. However, the engineering of a catalytic fusion protein nanopore is extremely challenging. Most reported biological nanopores are produced by oligomeric assembly from identical protein monomers. This has raised a significant bioengineering difficulty when a geometric symmetry between the enzyme and the pore has to be achieved [139]. Alternatively, a single helicase may be engineered to have hydrophobic outer surface mimicking a channel protein. However, this approach normally requires extremely specialized bioengineering efforts.

Alternatively, small enzyme proteins, serving as internal enzymatic adaptors, may be electrophoretically translocated through or trapped inside a large biological nanopore, such as ClyA (Figure 1.6C) [140]. Single-molecule kinetics of ligand binding to DHFR or AlkB can be seen in direct nanopore readouts. Stepwise current fluctuations, which may result from minute structural variations caused by ligand binding or dissociation to the internal enzymatic adaptor, have been observed in such a readout. This is extremely useful as a biomimetic approach in the design of single-molecule nanopore sensors for small molecules, with which the high selectivity and the efficiency of a natural enzyme has been utilized [136]. However, ClyA only accommodates enzymes with a highly compatible dimension. Trapping of the protein with its catalytic function retained may also be affected by the charge, size, or the amino acid distribution of the pore lumen [45]. It is also unclear whether the reported step-wise events result from change of conformation, orientation, or charge of the enzyme or a combination of these.

1.4 Summary and Prospects

In this chapter, the general methodology of biological nanopores has been demonstrated along with a variety of single-molecule sensing applications. Due to the limited length of this chapter, other emerging nanopore technologies, such as solid-state nanopores, nanopipettes, hybrid nanopores, or DNA nanopores, have not been thoroughly discussed. Biological nanopores, which originate from structural biology investigations of channel proteins, are extremely reliable protein templates with an atomic precision and consistency. This has equipped the biological nanopore with an exceptional sensing specificity, spatial resolution, and design precision. With the success of the Minion sequencer commercialized by Oxford Nanopore Technologies, genomic sequencing can now be carried out effectively with a hand-sized device. On the other hand, in addition to sequencing, a variety of unique applications have also been developed, including small analyte sensing, single-molecule chemistry, or single-molecule enzymology. Unlike most other single-molecule tools, which require delicate, complicated, and expensive instruments to achieve a single-molecule resolution, nanopore sensing is relatively simple and can be performed with much less costly apparatus. Thus, readers who are interested in carrying out an actual nanopore measurement may follow the highly simplified lab-protocol provided at the end of the chapter.

The methods discussed above are not without limitations, such as a limited measurement throughput, fragile lipid bilayers, a restricted pore size, and failure of pore assembly after heavy mutations. These may be overcome by emerging techniques such as fluorescence microscopy-based nanopore sensing [141–143], artificial membranes self-assembled from tri-block copolymers [144], and future development of protein engineering techniques [55]. One can also foresee the emergence of a variety of low-cost, disposable, and instrument-free point of care sensor chips [145], useful in clinical diagnosis.

Appendix: A Tutorial Protocol: Single-Molecule Discrimination of Different Cyclodextrin Types Using WT α-HL Nanopore

Purpose

Nanopore provides excellent resolution, and can immediately discriminate between molecular analytes with subtle differences in their chemical structures. Cyclodextrin (CD), which is an cyclic oligosaccharides consisting of 6, 7, or 8 glucopyranose units, has been discovered to have high binding affinities in the restriction of a WT α-HL nanopore.

Using cyclodextrin as the model analyte, the following tutorial presents the core procedures of nanopore sensing. Upon completion of the measurement, the readers will be able to visualize directly how different cyclodextrin molecules that differ only by one methyl group in each glucopyranose subunit could be directly recognized from single-channel recording results. Based on these demonstrations, the readers are strongly encouraged to custom design their own nanopore measurements subsequently.

Materials

- **Equipment**
  Electric arc generator (custom-made)
  Stereo microscope (JSZ5, Nanjing Jiangnan Novel Optics Co., Ltd)
  Patch-clamp amplifier (Axopatch 200B, Molecular Devices)
  Digitizer (Digidata 1550A1, Molecular Devices)
  Floating table (ZDT20-10, Jiangxi Liangsheng Technology Co., Ltd)
- **Consumables**
  Silver wire (1.5 mm in diameter, Alfa Aesar)
  Electrical wire (~1.5 mm in diameter, ~7 cm in length)
  Electrical connector (AMPLIMITE HDP-20, TE Connectivity)
  Measurement device and holder (custom-made)
  Teflon film (20 μm in thickness)
  Silicone Glue (3140RTV, DOW CORNING)
  Glass Capillary (0.3 mm in inner diameter, 100 mm in length)

Glass Capillary (1 mm in inner diameter, 100 mm in length)
Faraday Cage (custom-made)

- **Reagents**
  Sodium hypochlorite (NaOCl, OKA)
  Ultrapure water (Milli-Q)
  Hexadecane (Sigma-Aldrich)
  Pentane (Sigma-Aldrich)
  Potassium chloride (KCl, Aladdin)
  4-(2-Hydroxyethyl)-1-piperazine-ethanesulfonic acid (HEPES, Shanghai Yuanye Biotechnology)
  1,2-diphytanoyl-sn-glycero-3-phosphocholine (DPhPC, Avanti Polar Lipids)
  α-HL nanopores from *Escherichia coli* (custom-made)
  Trimethyl β-cyclodextrin (TM-βCD, Tokyo Chemical Industry)
  2,6-Dimethyl β-cyclodextrin (DM-βCD, Acmec)

Safety and Precautions

1. α-HL nanopores, also known as alpha-toxin, have been shown to be lethal in animals, causing respiratory paralysis, vascular and smooth muscle spasms, and tissue necrosis. Wear appropriate gloves and lab coat during manipulation.
2. Pentane is highly flammable. Its vapor and air can form an explosive mixture in case of high heat and open fire. Wear appropriate gloves and lab coat and handle in a fume hood.
3. Sodium hypochlorite is corrosives, unstable, and produces corrosive gases in case of high heat or solar radiation. Wear appropriate gloves and lab coat and handle in a fume hood.

Methods

- **Step 1: Preparation**

  1.1 **Preparation of Ag/AgCl electrode:** A pair of Ag/AgCl electrodes, which electrically connect the patch-clamp amplifier and electrolyte buffers in the measurement chambers, were custom made. A silver wire (1.5 mm in diameter, ~2 cm in length) and an electrical connector were soldered on each end of the electrical wire (Figure 1.7A). To form a uniform AgCl coating, the silver wire part of the electrode was immersed in 30% (v/v) NaOCl (diluted in water) for 12 h. After that, the silver wire part of the electrode was thoroughly rinsed with ultrapure water and air-dried with nitrogen gas prior to use.

1.2 **Construction of measurement device:** The measurement device is composed of two custom polyformaldehyde modules (Figure 1.7B) separated by a ~20 μm thick Teflon film with an aperture. The aperture on the Teflon film serves as a solid support for lipid bilayer assembly. To custom make the measurement device, an aperture with a desired geometry (~100 μm in diameter) was generated using a custom-made electric arc generator, and subsequently characterized by stereomicroscopy (Figure 1.7C). Subsequently, the Teflon film was clamped in the middle of the two polyformaldehyde modules and sealed with silicone glue. The device was left untouched for 12 h until the glue was set completely.

Cautious: *The aperture size has a direct influence on the bilayer stability. The desired diameter of the aperture is 50 to 150 μm.*

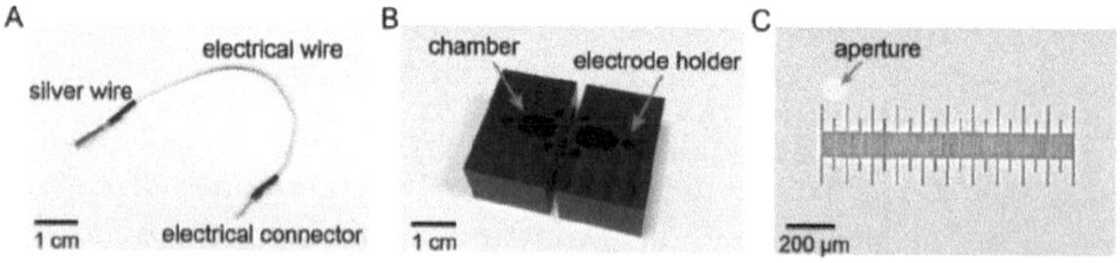

Figure 1.7 Measurement components for a single-channel recording. (A) Photograph of a Ag/AgCl electrode. (B) Photograph of a measurement device. Each module of the device has a 500 μL chamber and three electrode holders. (C) Microscopic characterization of the aperture on the Teflon film using a stereo microscope. The desired aperture has a round geometry with a smooth edge and measures ~100 μm in diameter.

- **Step 2: Electrophysiology Recordings**

2.1 **Tapping hexadecane on the aperture:** Hexadecane helps lipids to self-assemble in the aperture. To achieve a uniform dispersion of hexadecane around the aperture, 0.5% (v/v) hexadecane (dissolved in pentane) was siphoned into a glass capillary (0.3 mm in inner diameter) and tapped on both sides of the aperture. Fast evaporation of pentane was observed, quickly resulting in a uniform dispersion of hexadecane.

2.2 **Measurement setup:** All electrophysiology measurements were performed with an Axonpatch 200B patch-clamp amplifier paired with a Digidata 1550A1 digitizer. Initially, the measurement device was fixed on a custom-made holder. A pair of Ag/AgCl electrodes, which were electrically connected to the patch-clamp amplifier, were inserted in electrode holders in opposing chambers respectively. Conventionally, the chamber which is electrically grounded was defined as *Cis*, while the other chamber was defined as *Trans*.

Cautions: *To avoid external electrical noise, the device should be electrically shielded in a custom Faraday cage (34 cm × 23 cm × 15 cm) during the measurement. To avoid mechanical vibrations, the Faraday cage should be mounted on a floating table during the measurement.*

2.3 **Lipid bilayer formation:** Both chambers of the device are filled with 500 μL of electrolyte buffers (1.5 M KCl, 10 mM HEPES, pH 7.0), which forms a direct electrochemical connection with the Ag/AgCl electrodes to form a circuit. Without a lipid bilayer sealing the aperture on the Teflon film, the measured signal at this moment reports an "overload" current. To form the lipid bilayer, 100 μL pentane solution of DPhPC (5 mg/mL) was initially dropped in both chambers using a glass capillary (1 mm in inner diameter). Subsequently, by repetitively pipetting the electrolyte buffer in the *Cis* or the *Trans* chamber up and down until the measured current reverts from "overload" to ~0 pA, a self-assembled lipid bilayer is formed.

2.4 **Bilayer characterization:** To exclude the possibility of an air bubble trapping or leaky bilayer formation, the bilayer must be characterized before pore insertion. By applying a triangle wave voltage protocol (peak amplitude: ±50 mV, frequency: 5 Hz, Figure 1.8A, top), the bilayer quality could be judged from the shape of the current readout (Figure 1.8A, bottom). A well-formed, nonleaky bilayer normally reports a square wave-shaped current with a flat plateau (Figure 1.8A, bottom). The bilayer capacitance can be determined according to $C = \dfrac{I}{dV/dt}$, where I stands for the absolute current magnitude of the square wave signal. A leak-free bilayer with a capacitance of 50–100 pF is desired for pore insertion.

2.5 **Nanopore insertion:** Initially, α-HL nanopore (~0.2 ng) was added to the *Cis* chamber for spontaneous pore insertion. To stimulate pore insertion, a high amplitude, square wave-shaped voltage protocol (amplitude: 200 mV, frequency: 1 Hz) was applied until an abrupt current jump of ~300 pA is observed (Figure 1.8B), which marks the insertion of a single α-HL pore in the bilayer was observed. Subsequently, the *Cis* chamber was exchanged with fresh electrolyte buffer to avoid further pore insertions.

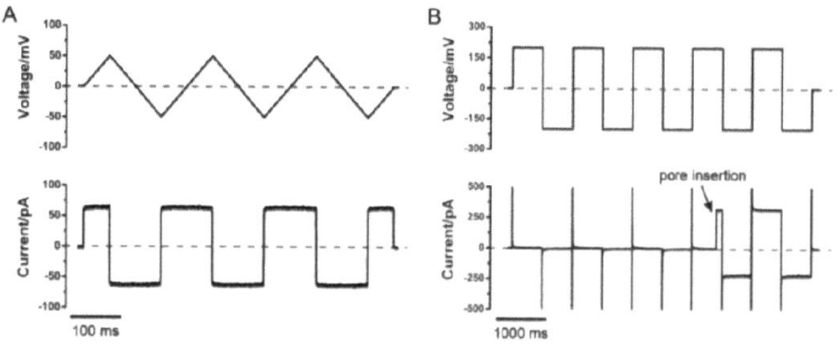

Figure 1.8 Bilayer characterization and nanopore insertion. (A) Bilayer capacitance measurement. To estimate the bilayer capacitance, a triangular wave voltage protocol was applied (peak amplitude: ±50 mV, frequency: 5 Hz) to produce a square wave current output. The bilayer capacitance was derived according to $C = \frac{I}{dV/dt}$, where I stands for the amplitude of the current plateau of the square wave-shaped current signal. (B) α-HL nanopore insertion. A square wave voltage protocol (amplitude: 200 mV, frequency: 1 Hz) was applied to stimulate pore insertion. A single nanopore insertion was observed from an abrupt current increase with the predicted current amplitude.

- **Step 3: Single-Molecule Sensing**

3.1 Background recordings: By continuously applying a +20 mV voltage bias, the current output was recorded with a sampling rate of 25 kHz and low-pass filtered at 5 kHz (Figure 1.9A). As previously reported, the open pore current of an α-HL nanopore measures ~30 pA at +20 mV with a 1.5 M KCl buffer. In the absence of analytes, single-channel recording from a single, well-assembled α-HL nanopore produces clear background signals with no obvious gating or resistive pulse events.

3.2 Sensing TM-βCD: About 0.5 μL TM-βCD (100 mM stock solution, dissolved in ultrapure water) was added to the

Trans chamber and magnetic stirred. By continuously applying a voltage bias of +20 mV, the current was recorded with a sampling rate of 25 kHz and low-pass filtered at 5 kHz (Figure 1.9B). Binding of TM-βCD with an α-HL nanopore produces long residing and deep pore blockade signals.

3.3 **Simultaneous sensing TM-βCD and DM-βCD:** A further 0.1 μL of DM-βCD (100 mM stock solution, dissolved in ultrapure water) was added to the *Trans* chamber and stirred magnetically. The current was recorded with a sampling rate of 25 kHz and low-pass filtered at 5 kHz (Figure 1.9C). Two types of resistive pulses, which are different in the depth and duration of the blockage event, are now clearly observed. This signal difference results from a minor chemical structure variation between TM-βCD and DM-βCD, which forms the sensing basis of this experiment.

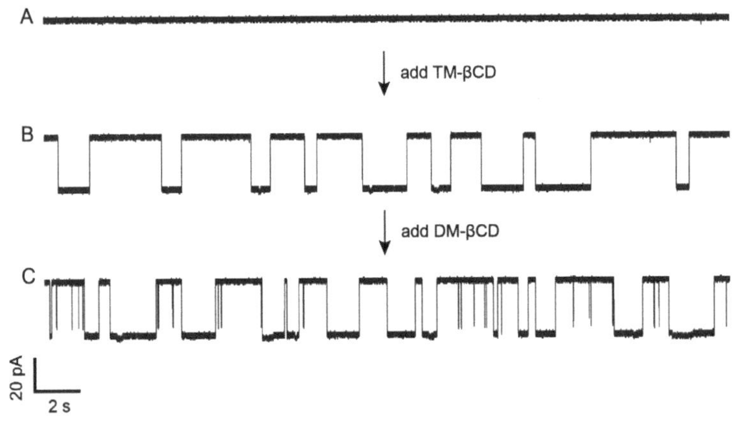

Figure 1.9 Stochastic sensing of different βCDs. (A) Background current. The current trace of a single α-HL nanopore shows no gating events in the absence of analytes. (B) TM-βCD sensing. Successive deep and long residing current blockades are clearly observed when TM-βCD is added to the *Trans* chamber with a 100 μM final concentration. (C) TM-βCD and DM-βCD sensing. By subsequently adding DM-βCD with a 20 μM final concentration to the *Trans* chamber, a new signal type (shallow and short residing events) start to appear. The measurement in (B and C) is carried out with a 1.5 M KCl buffer and an applied potential of +20 mV. The electrophysiology traces are recorded with a sampling rate of 25 kHz and low-pass filtered at 5 kHz.

- **Data Analysis**

Data analysis was performed with Clampfit 10.7 (Molecular Devices). The "single-channel research" option in Clampfit 10.7

was used to extract βCD translocation events. Characteristic parameters, such as blockade current (***I_b***), dwell time (***t_{off}***), interevent interval (***t_{on}***), are reported. TM-βCD and DM-βCD could be discriminated from their different blockade current (***I_b***) values (Figure 1.10A). A histogram of the (***I_b***) values reports two distinct event distributions. A separation of ~5 pA was estimated according to the Gaussian fitting result (Figure 1.10B).

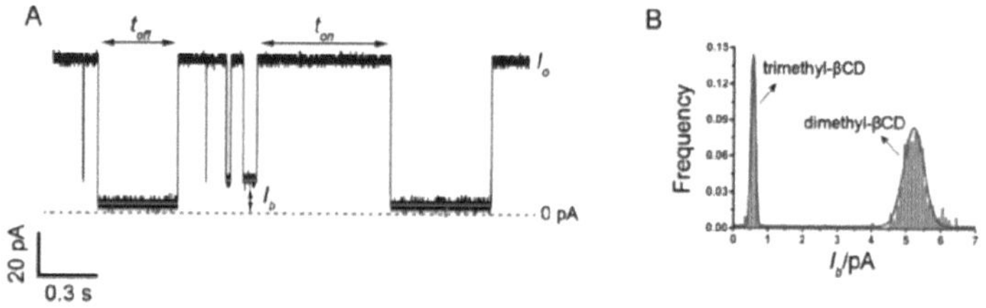

Figure 1.10 Event statistics. (A) A representative current trace acquired from simultaneous sensing of TM-βCD and DM-βCD using α-HL. The open pore current (I_o), blockade current (I_b), event dwell time (t_{off}), and the interevent interval (t_{on}) are defined as marked on the trace. Translocation events, which were extracted by ClampFit, are marked with red lines. (B) Histogram of the blockade events. Different βCD types are discriminated according to their distinct blockage depth (I_b).

References

1. Sigworth FJ, Neher E (1980). Single NA+ channel currents observed in cultured rat muscle-cells, *Nature,* **287**, 447–449.
2. Williams DBC, Carter CB (1996). *Transmission Electron Microscopy,* Springer, Boston, MA, pp 3–17.
3. Binnig G, Rohrer H, Gerber C, Weibel E (1982). Tunneling through a controllable vacuum gap, *Appl Phys Lett,* **40**, 178–180.
4. Ashkin A, Dziedzic, JM (1987). Optical trapping and manipulation of viruses and bacteria, *Science* **235**, 1517–1520.
5. Strick TR, Allemand JF, Bensimon D, Bensimon A, Croquette V (1996). The elasticity of a single supercoiled DNA molecule. *Science,* **271**, 1835–1837.
6. Peterman EJG, Sosa H, Moerner WE (2004). Single-molecule fluorescence spectroscopy and microscopy of biomolecular motors, *Annu Rev Phys Chem,* **55**, 79–96.

7. Hell SW, Wichmann J (1994). Breaking the diffraction resolution limit by stimulated-emission—Stimulated-emission-depletion fluorescence microscopy, *Opt Lett*, **19**, 780–782.

8. Ha T, et al. (1996). Probing the interaction between two single molecules: Fluorescence resonance energy transfer between a single donor and a single acceptor, *Proc Natl Acad Sci USA*, **93**, 6264–6268.

9. Tokunaga M, Kitamura K, Saito K, Iwane AH, Yanagida T (1997). Single molecule imaging of fluorophores and enzymatic reactions achieved by objective-type total internal reflection fluorescence microscopy, *Biochem Biophys Res Commun*, **235**, 47–53.

10. van Oijen AM, et al. (2003). Single-molecule kinetics of lambda exonuclease reveal base dependence and dynamic disorder, *Science*, **301**, 1235–1238.

11. Lu HP, Xun L, Xie XS (1998). Single-molecule enzymatic dynamics, *Science*, **282**, 1877–1882.

12. Schuler B, Lipman EA, Eaton WA (2002). Probing the free-energy surface for protein folding with single-molecule fluorescence spectroscopy, *Nature*, **419**, 743–747.

13. Cui XD, et al. (2001). Reproducible measurement of single-molecule conductivity, *Science* **294**, 571–574.

14. Benoit M, Gabriel D, Gerisch G, Gaub HE (2000). Discrete interactions in cell adhesion measured by single-molecule force spectroscopy, *Nat Cell Biol*, **2**, 313–317.

15. Kawano R, Schibel AEP, Cauley C, White HS (2009). Controlling the translocation of single-stranded DNA through alpha-hemolysin ion channels using viscosity, *Langmuir*, **25**, 1233–1237.

16. Song L, et al. (1996). Structure of staphylococcal alpha-hemolysin, a heptameric transmembrane pore, *Science*, **274**, 1859–1866.

17. Bull BS, Schneiderman MA, Brecher G (1965). Platelet counts with the Coulter counter. *Am J Clin Pathol*, **44**, 678–688.

18. Akeson M, Branton D, Kasianowicz JJ, Brandin E, Deamer DW (1999). Microsecond time-scale discrimination among polycytidylic acid, polyadenylic acid, and polyuridylic acid as homopolymers or as segments within single RNA molecules, *Biophys J*, **77**, 3227–3233.

19. Kasianowicz JJ, Brandin E, Branton D, Deamer DW (1996). Characterization of individual polynucleotide molecules using a membrane channel, *Proc Natl Acad Sci USA*, **93**, 13770–13773.

20. Branton D, et al. (2008). The potential and challenges of nanopore sequencing, *Nat Biotechnol*, **26**, 1146–1153.

21. Rodrigues CG, Machado DC, Chevtchenko SF, Krasilnikov OV (2008). Mechanism of KCl enhancement in detection of nonionic polymers by nanopore sensors, *Biophys J*, **95**, 5186–5192.
22. Kang XF, Gu LQ, Cheley S, Bayley H (2005). Single protein pores containing molecular adapters at high temperatures. *Angew Chem Int Edit*, **44**, 1495–1499.
23. Haque F, Li J, Wu HC, Liang XJ, Guo P (2013). Solid-state and biological nanopore for real-time sensing of single chemical and sequencing of DNA, *Nano Today*, **8**, 56–74.
24. Pastoriza-Gallego M, et al. (2011). Dynamics of unfolded protein transport through an aerolysin pore, *J Am Chem Soc*, **133**, 2923–2931.
25. Pastoriza-Gallego M, et al. (2007). Urea denaturation of alpha-hemolysin pore inserted in planar lipid bilayer detected by single nanopore recording: Loss of structural asymmetry, *Febs Lett*, **581**, 3371–3376.
26. Oukhaled G, et al. (2007). Unfolding of proteins and long transient conformations detected by single nanopore recording. *Phys Rev Lett*, **98**, 158101.
27. Li J, et al. (2001). Ion-beam sculpting at nanometre length scales. *Nature*, **412**, 166–169.
28. Storm AJ, Chen JH, Ling XS, Zandbergen HW, Dekker C (2003). Fabrication of solid-state nanopores with single-nanometre precision, *Nat Mater*, **2**, 537–540.
29. Siwy Z, Fulinski A (2002). Fabrication of a synthetic nanopore ion pump, *Phys Rev Lett*, **89**, 198103.
30. Kwok H, Briggs K, Tabard-Cossa V (2014). Nanopore fabrication by controlled dielectric breakdown, *Plos One*, **9**, e92880.
31. Yusko EC, et al. (2011). Controlling protein translocation through nanopores with bio-inspired fluid walls, *Nat Nanotechnol*, **6**, 253–260.
32. Hall AR, et al. (2010). Hybrid pore formation by directed insertion of alpha-haemolysin into solid-state nanopores, *Nat Nanotechnol*, **5**, 874–877.
33. Wei RS, Martin TG, Rant U, Dietz H (2012). DNA origami gatekeepers for solid-state nanopores, *Angew Chem Int Edit*, **51**, 4864–4867.
34. Bell NAW, et al. (2012). DNA origami nanopores, *Nano Lett*, **12**, 512–517.
35. Langecker M, et al. (2012). Synthetic lipid membrane channels formed by designed DNA nanostructures, *Science*, **338**, 932–936.

36. Butler TZ, Pavlenok M, Derrington IM, Niederweis M, Gundlach JH (2008). Single-molecule DNA detection with an engineered MspA protein nanopore, *Proc Natl Acad Sci USA,* **105**, 20647–20652.

37. Thakur AK, Movileanu L (2019). Real-time measurement of protein-protein interactions at single-molecule resolution using a biological nanopore, *Nat Biotechnol,* **37**, 96–101.

38. Jing P, Haque F, Vonderheide AP, Montemagno C, Guo PX (2010). Robust properties of membrane-embedded connector channel of bacterial virus phi29 DNA packaging motor, *Mol Biosyst,* **6**, 1844–1852.

39. Soskine M, et al. (2012). An engineered ClyA nanopore detects folded target proteins by selective external association and pore entry, *Nano Lett,* **12**, 4895–4900.

40. Fahie, MA, Yang B, Mullis M, Holden MA, Chen M (2015). Selective detection of protein homologues in serum using an OmpG nanopore, *Analyt Chem,* **87**, 11143–11149.

41. Ionescu SA, et al. (2017). Orientation of the OmpF porin in planar lipid bilayers, *Chembiochem,* **18**, 554–562.

42. Tanaka K, Caaveiro JMM, Morante K, Manuel Gonzalez-Manas J, Tsumoto K (2015). Structural basis for self-assembly of a cytolytic pore lined by protein and lipid, *Nat Commun,* **6,** 6337.

43. Cao BH, et al. (2014). Structure of the nonameric bacterial amyloid secretion channel, *Proc Natl Acad Sci USA,* **111**, E5439–E5444.

44. Wang H-Y, et al. (2013). Single-molecule DNA detection using a novel SP1 protein nanopore, *Chem Commun,* **49**, 1741–1743.

45. Franceschini L, Brouns T, Willems K, Carlon E, Maglia G (2016). DNA translocation through nanopores at physiological ionic strengths requires precise nanoscale engineering, *ACS Nano,* **10**, 8394–8402.

46. Stoddart D, Heron AJ, Mikhailova E, Maglia G, Bayley H (2009). Single-nucleotide discrimination in immobilized DNA oligonucleotides with a biological nanopore, *Proc Natl Acad Sci USA,* **106**, 7702–7707.

47. Faller M, Niederweis M, Schulz GE (2004). The structure of a mycobacterial outer-membrane channel, *Science,* **303**, 1189–1192.

48. Locher KP, et al. (1998). Transmembrane signaling across the ligand-gated FhuA receptor: Crystal structures of free and ferrichrome-bound states reveal allosteric changes, *Cell,* **95**, 771–778.

49. Subbarao GV, van den Berg B (2006). Crystal structure of the monomeric porin OmpG, *J Mol Biol,* **360**, 750–759.

50. Iacovache I, et al. (2016). Cryo-EM structure of aerolysin variants reveals a novel protein fold and the pore-formation process, *Nat Commun,* **7,** 12062.

51. Cowan SW, et al. (1995). The structure of OmpF porin in a tetragonal crystal form, *Structure,* **3**, 1041–1050.

52. Simpson AA, et al. (2000). Structure of the bacteriophage phi 29 DNA packaging motor, *Nature,* **408**, 745–750.

53. Peng W, Santos MdS, Li Y, Tomchick DR, Orth K (2019). High-resolution cryo-EM structures of the *E. coli* hemolysin ClyA oligomers, *PLoS One,* **14,** e0213423.

54. Wang Y, et al. (2018). Osmosis-driven motion-type modulation of biological nanopores for parallel optical nucleic acid sensing, *ACS Appl Mater Interfaces,* **10**, 7788–7797.

55. Ayub, M, Bayley H (2016). Engineered transmembrane pores, *Curr Opin Chem Biol,* **34**, 117–126.

56. Rosenstein JK, Wanunu M, Merchant CA, Drndic M, Shepard KL (2012). Integrated nanopore sensing platform with sub-microsecond temporal resolution, *Nat Methods,* **9**, 487–U112.

57. Huang S (2014). Nanopore-based sensing devices and applications to genome sequencing: A brief history and the missing pieces, *Chin Sci Bull,* **59**, 4918–4928.

58. Lee M-H, et al. (2014). A low-noise solid-state nanopore platform based on a highly insulating substrate, *Sci Rep-UK,* **4,** 7448.

59. Maglia G, Heron AJ, Stoddart D, Japrung D, Bayley H (2010). In: NG Walter (eds) *Methods in Enzymology*, Vol 475: *Single Molecule Tools, Pt B: Super-Resolution, Particle Tracking, Multiparameter, and Force Based Methods*, vol. 475, pp 591–623.

60. Manrao EA, et al. (2012). Reading DNA at single-nucleotide resolution with a mutant MspA nanopore and phi29 DNA polymerase, *Nat Biotechnol,* **30**, 349–U174.

61. Sauer-Budge AF, Nyamwanda JA, Lubensky DK, Branton D (2003). Unzipping kinetics of double-stranded DNA in a nanopore, *Phys Rev Lett,* **90,** 238101.

62. Nivala J, Marks DB, Akeson M (2013). Unfoldase-mediated protein translocation through an alpha-hemolysin nanopore, *Nat Biotechnol,* **31**, 247–250.

63. Meller A, Nivon L, Brandin E, Golovchenko J, Branton D (2000). Rapid nanopore discrimination between single polynucleotide molecules, *Proc Natl Acad Sci USA,* **97**, 1079–1084.

64. Sanchez-Quesada J, Saghatelian A, Cheley S, Bayley H, Ghadiri MR (2004). Single DNA rotaxanes of a transmembrane pore protein, *Angew Chem Int Edit,* **43**, 3063–3067.

65. Purnell RF, Schmidt JJ (2009). Discrimination of single base substitutions in a DNA strand immobilized in a biological nanopore, *ACS Nano,* **3**, 2533–2538.

66. Wallace EV, et al. (2010). Identification of epigenetic DNA modifications with a protein nanopore, *Chem Commun,* **46**, 8195–8197.

67. Ayub M, Bayley H (2012). Individual RNA base recognition in immobilized oligonucleotides using a protein nanopore, *Nano Lett,* **12**, 5637–5643.

68. Manrao EA, Derrington IM, Pavlenok M, Niederweis M, Gundlach JH (2011). Nucleotide discrimination with DNA immobilized in the MspA nanopore, *PLoS One,* **6**, e25723.

69. Meller A, Branton D (2002). Single molecule measurements of DNA transport through a nanopore, *Electrophoresis,* **23**, 2583–2591.

70. Meller A, Nivon L, Branton D (2001). Voltage-driven DNA translocations through a nanopore, *Phys Rev Lett,* **86**, 3435–3438.

71. Mitchell N, Howorka S (2008). Chemical tags facilitate the sensing of individual DNA strands with nanopores, *Angew Chem Int Edit,* **47**, 5565–5568.

72. de Zoysa RSS, et al. (2009). Slowing DNA translocation through nanopores using a solution containing organic salts, *J Phys Chem B,* **113**, 13332–13336.

73. Rincon-Restrepo M, Milthallova E, Bayley H, Maglia G (2011). Controlled translocation of individual DNA molecules through protein nanopores with engineered molecular brakes, *Nano Lett,* **11**, 746–750.

74. Lu H, Giordano F, Ning Z (2016). Oxford Nanopore MinION sequencing and genome assembly, *Genom Proteom Bioinf,* **14**, 265–279.

75. Jain M, Olsen HE, Paten B, Akeson M (2016). The Oxford Nanopore MinION: Delivery of nanopore sequencing to the genomics community, *Genom Biol,* **17**, 239.

76. Lieberman KR, et al. (2010). Processive replication of single DNA molecules in a nanopore catalyzed by phi29 DNA polymerase, *J Am Chem Soc,* **132**, 17961–17972.

77. Cherf GM, et al. (2012). Automated forward and reverse ratcheting of DNA in a nanopore at 5-A precision, *Nat Biotechnol,* **30**, 344–348.

78. Laszlo AH, et al. (2014). Decoding long nanopore sequencing reads of natural DNA, *Nat Biotechnol,* **32**, 829–833.

79. Wescoe ZL, Schreiber J, Akeson M (2014). Nanopores discriminate among five C5-cytosine variants in DNA, *J Am Chem Soc,* **136**, 16582–16587.

80. Derrington IM, et al. (2015). Subangstrom single-molecule measurements of motor proteins using a nanopore, *Nat Biotechnol,* **33**, 1073–1075.

81. Eisenstein M (2012). Oxford nanopore announcement sets sequencing sector abuzz, *Nat Biotechnol,* **30**, 295–296.

82. Goodwin S, et al. (2015). Oxford Nanopore sequencing, hybrid error correction, and de novo assembly of a eukaryotic genome, *Genom Res,* **25**, 1750–1756.

83. Loman NJ, Quick J, Simpson JT (2015). A complete bacterial genome assembled de novo using only nanopore sequencing data, *Nat Methods,* **12**, 733–U751.

84. Quick J, et al. (2016). Real-time, portable genome sequencing for Ebola surveillance, *Nature,* **530**, 228–232.

85. Astier Y, Braha O, Bayley H (2006). Toward single molecule DNA sequencing: Direct identification of ribonucleoside and deoxyribonucleoside 5 '-monophosphates by using an engineered protein nanopore equipped with a molecular adapter, *J Am Chem Soc,* **128**, 1705–1710.

86. Wu HC, Astier Y, Maglia G, Mikhailova E, Bayley H (2007). Protein nanopores with covalently attached molecular adapters, *J Am Chem Soc,* **129**, 16142–16148.

87. Clarke J, et al. (2009). Continuous base identification for single-molecule nanopore DNA sequencing, *Nat Nanotechnol,* **4**, 265–270.

88. Ayub M, Hardwick SW, Luisi BF, Bayley H (2013). Nanopore-based identification of individual nucleotides for direct RNA sequencing, *Nano Lett,* **13**, 6144–6150.

89. Kumar S, et al. (2012). PEG-labeled nucleotides and nanopore detection for single molecule DNA sequencing by synthesis, *Sci Rep-UK,* **2,** 684.

90. Fuller CW, et al. (2016). Real-time single-molecule electronic DNA sequencing by synthesis using polymer-tagged nucleotides on a nanopore array, *Proc Natl Acad Sci USA,* **113**, 5233–5238.

91. Restrepo-Perez L, Joo C, Dekker C (2018). Paving the way to single-molecule protein sequencing, *Nat Nanotechnol,* **13**, 786–796.

92. Sutherland TC, et al. (2004). Structure of peptides investigated by nanopore analysis, *Nano Lett*, **4**, 1273–1277.

93. Movileanu L, Schmittschmitt JP, Scholtz JM, Bayley H (2005). Interactions of peptides with a protein pore, *Biophys J*, **89**, 1030–1045.

94. Goodrich CP, et al. (2007). Single-molecule electrophoresis of beta-hairpin peptides by electrical recordings and Langevin dynamics simulations, *J Phys Chem B*, **111**, 3332–3335.

95. Mohammad MM, Movileanu L (2008). Excursion of a single polypeptide into a protein pore: Simple physics, but complicated biology, *Eur Biophys J Biophys Lett*, **37**, 913–925.

96. Merstorf C, et al. (2012). Wild type, mutant protein unfolding and phase transition detected by single-nanopore recording, *ACS Chem Biol*, **7**, 652–658.

97. Kennedy E, Dong Z, Tennant C, Timp G (2016). Reading the primary structure of a protein with 0.07 nm(3) resolution using a subnanometre-diameter pore, *Nat Nanotechnol*, **11**, 968–976.

98. Dong Z, Kennedy E, Hokmabadi M, Timp G (2017). Discriminating residue substitutions in a single protein molecule using a sub-nanopore, *ACS Nano*, **11**, 5440–5452.

99. Payet L, et al. (2012). Thermal unfolding of proteins probed at the single molecule level using nanopores, *Anal Chem*, **84**, 4071–4076.

100. Rodriguez-Larrea D, Bayley H (2013). Multistep protein unfolding during nanopore translocation, *Nat Nanotechnol*, **8**, 288–295.

101. Kawano R, et al. (2011). Rapid detection of a cocaine-binding aptamer using biological nanopores on a chip, *J Am Chem Soc*, **133**, 8474–8477.

102. Gu LQ, Braha O, Conlan S, Cheley S, Bayley H (1999). Stochastic sensing of organic analytes by a pore-forming protein containing a molecular adapter, *Nature*, **398**, 686–690.

103. Guan X, Gu LQ, Cheley S, Braha O, Bayley H (2005). Stochastic sensing of TNT with a genetically engineered pore, *Chembiochem*, **6**, 1875–1881.

104. Kang XF, Cheley S, Guan XY, Bayley H (2006). Stochastic detection of enantiomers, *J Am Chem Soc*, **128**, 10684–10685.

105. Wang D, Zhao Q, Zoysa RSSd, Guan X (2009). Detection of nerve agent hydrolytes in an engineered nanopore, *Sens Actuators B Chem*, **139**, 440–446.

106. Boersma AJ, Brain KL, Bayley H (2012). Real-time stochastic detection of multiple neurotransmitters with a protein nanopore, *ACS Nano*, **6**, 5304–5308.

107. Haque F, Lunn J, Fang HM, Smithrud D, Guo PX (2012). Real-time sensing and discrimination of single chemicals using the channel of Phi29 DNA packaging nanomotor, *ACS Nano*, **6**, 3251–3261.

108. Gupta J, Zhao Q, Wang G, Kang X, Guan X (2013). Simultaneous detection of CMPA and PMPA, hydrolytes of soman and cyclosarin nerve agents, by nanopore analysis, *Sens Actuators B Chem*, **176**, 625–631.

109. Cheley S, Gu LQ, Bayley H (2002). Stochastic sensing of nanomolar inositol 1,4,5-trisphosphate with an engineered pore, *Chem Biol*, **9**, 829–838.

110. Braha O, et al (2000). Simultaneous stochastic sensing of divalent metal ions, *Nat Biotechnol*, **18**, 1005–1007.

111. Wen S, et al. (2011). Highly sensitive and selective DNA-based detection of mercury(II) with alpha-hemolysin nanopore, *J Am Chem Soc*, **133**, 18312–18317.

112. Wang G, Zhao Q, Kang X, Guan X (2013). Probing mercury(II)-DNA interactions by nanopore stochastic sensing, *J Phys Chem B*, **117**, 4763–4769.

113. Yang C, et al. (2013). Highly sensitive simultaneous detection of lead(II) and barium(II) with G-quadruplex DNA in alpha-hemolysin nanopore, *Analyt Chem*, **85**, 7302–7307.

114. Braha O, et al. (1997). Designed protein pores as components for biosensors, *Chem Biol*, **4**, 497–505.

115. Cao J, et al. (2019). Giant single molecule chemistry events observed from a tetrachloroaurate(III) embedded *Mycobacterium smegmatis* porin A nanopore, *Nat Commun*, **10**, 5668.

116. Qing Y, Pulcu GS, Bell NAW, Bayley H (2018). Bioorthogonal cycloadditions with sub-millisecond intermediates, *Angew Chem*, **57**, 1218–1221.

117. Shin SH, Luchian T, Cheley S, Braha O, Bayley H (2002). Kinetics of a reversible covalent-bond-forming reaction observed at the single-molecule level, *Angew Chem Int Edit*, **41**, 3707–3709.

118. Steffensen MB, Rotem D, Bayley H (2014). Single-molecule analysis of chirality in a multicomponent reaction network, *Nat Chem*, **6**, 604–608.

119. Pulcu GS, Mikhailova E, Choi LS, Bayley H (2015). Continuous observation of the stochastic motion of an individual small-molecule walker, *Nat Nanotechnol*, **10**, 76–83.

120. Ramsay WJ, Bell NAW, Qing Y, Bayley H (2018). Single-molecule observation of the intermediates in a catalytic cycle, *J Am Chem Soc,* **140**, 17538–17546.

121. Shin SH, Bayley H (2005). Stepwise growth of a single polymer chain, *J Am Chem Soc,* **127**, 10462–10463.

122. Loudwig S, Bayley H (2006). Photoisomerization of an individual azobenzene molecule in water: An on-off switch triggered by light at a fixed wavelength, *J Am Chem Soc,* **128**, 12404–12405.

123. Ramsay WJ, Bayley H (2018). Single-molecule determination of the isomers of D-glucose and D-fructose that bind to boronic acids, *Angew Chem Int Edit,* **57**, 2841–2845.

124. Lee J, Bayley H (2015). Semisynthetic protein nanoreactor for single-molecule chemistry, *Proc Natl Acad Sci USA,* **112**, 13768–13773.

125. Lee J, et al. (2016). Semisynthetic nanoreactor for reversible single-molecule covalent chemistry, *ACS Nano,* **10**, 8843–8850.

126. Ha T, Kozlov AG, Lohman TM (2012). Single-molecule views of protein movement on single-stranded DNA, *Annu Rev Biophys,* **41**, 295–319.

127. Kim H, Ha T (2013). Single-molecule nanometry for biological physics, *Rep Prog Phys,* **76,** 016601.

128. Dulin D, et al. (2015). High spatiotemporal-resolution magnetic tweezers: Calibration and applications for DNA dynamics, *Biophys J,* **109**, 2113–2125.

129. Meyer G, Amer NM (1988). Novel optical approach to atomic force microscopy, *Appl Phys Lett,* **53**, 1045–1047.

130. Lohman TM, Tomko EJ, Wu CG (2008). Non-hexameric DNA helicases and translocases: Mechanisms and regulation, *Nat Rev Mol Cell Bio,* **9**, 391–401.

131. Byrd AK, Raney KD (2012). Superfamily 2 helicases, *Front Biosci-Landmrk,* **17**, 2070–2088.

132. Hornblower B, et al. (2007). Single-molecule analysis of DNA-protein complexes using nanopores, *Nat Methods,* **4**, 315–317.

133. Wilson NA, et al. (2009). Electronic control of DNA polymerase binding and unbinding to single DNA molecules, *ACS Nano,* **3**, 995–1003.

134. Olasagasti F, et al. (2010). Replication of individual DNA molecules under electronic control using a protein nanopore, *Nat Nanotechnol,* **5**, 798–806.

135. Laszlo AH, Derrington IM, Gundlach JH (2016). MspA nanopore as a single-molecule tool: From sequencing to SPRNT, *Methods,* **105**, 75–89.
136. Galenkamp NS, Soskine M, Hermans J, Wloka C, Maglia G (2018). Direct electrical quantification of glucose and asparagine from bodily fluids using nanopores, *Nat Commun,* **9,** 4085.
137. Craig JM, et al. (2019). Determining the effects of DNA sequence on Hel308 helicase translocation along single-stranded DNA using nanopore tweezers, *Nucleic Acids Res,* **47**, 2506–2513.
138. Laszlo AH, Derrrington IM, Gundlach JH (2017). Subangstrom measurements of enzyme function using a biological nanopore, SPRNT, *Methods Enzymol,* **582**, 387–414.
139. Ho C-W, et al. (2015). Engineering a nanopore with co-chaperonin function, *Sc Adv,* **1,** e1500905.
140. Soskine M, Biesemans A, Maglia G (2015). Single-molecule analyte recognition with ClyA nanopores equipped with internal protein adaptors, *J Am Chem Soc,* **137**, 5793–5797.
141. Heron AJ, Thompson JR, Cronin B, Bayley H, Wallace MI (2009). Simultaneous measurement of ionic current and fluorescence from single protein pores, *J Am Chem Soc,* **131**, 1652–1653.
142. Huang S, Romero-Ruiz M, Castell OK, Bayley H, Wallace MI (2015). High-throughput optical sensing of nucleic acids in a nanopore array, *Nat Nanotechnol,* **10**, 986–U996.
143. Wang YQ, et al. (2018). Osmosis-driven motion-type modulation of biological nanopores for parallel optical nucleic acid sensing, *ACS Appl Mater Interfaces,* **10**, 7788–7797.
144. Lang C, et al. (2018). Creating cross-linked lamellar block copolymer supporting layers for biomimetic membranes, *Faraday Discuss,* **209**, 179–191.
145. Wang Y, et al. (2019). Electrode-free nanopore sensing by DiffusiOptoPhysiology, *Sci Adv,* **5,** eaar3309.

Chapter 2

Optical Tweezers for Manipulation of Single Molecules

Guangtao Song[a,b] and Yan Zeng[a]

[a]*Key Laboratory of RNA Biology, CAS Center for Excellence in Biomacromolecules, Institute of Biophysics, Chinese Academy of Sciences, Beijing 100101, China*

[b]*University of Chinese Academy of Sciences, Beijing 100049, China*

gsong@ibp.ac.cn

Optical tweezers, also known as an optical trap, use the forces exerted by a highly focused laser beam to trap and move micron-sized dielectric particles. Typically, optical tweezers can obtain a nanometer spatial resolution, a piconewton force resolution, and a millisecond time resolution, and thus they are excellently suited to studies of biological processes at the single-molecule level. Since the pioneering work of Arthur Ashkin, optical tweezers have become a versatile tool with which to address numerous biochemical and biophysical processes, from the basic mechanical properties of biomacromolecules such as DNA, RNA, and proteins to the multitude of molecular machines including, for example, RNA polymerase, molecular motors, and ribosomes that drive the internal dynamics of the cell. In this chapter, we first provide an introduction to the theory and design of optical tweezers. Then, we describe the optical

Single-Molecule Tools for Bioanalysis
Edited by Shuo Huang

ISBN 978-981-4800-44-0 (Hardcover), 978-1-003-18913-8 (eBook)
www.jennystanford.com

tweezers apparatus, including instrument design considerations, position detection schemes, and calibration techniques. We review some important progresses in single-molecule manipulation by the use of optical tweezers and finally, we provide a detailed protocol to study folding of DNA hairpin using optical tweezers.

2.1 Introduction

Force is the basis of a fundamental and regulatory mechanism of biomolecular interactions, which drive many cellular processes. For example, DNA and RNA polymerases generate forces during DNA replication and RNA transcription, and the microtubule motors kinesin and cytoplasmic dynein generate forces that regulate spindle and chromosome positioning during mitosis [1, 2]. The ability to apply external forces in a precisely defined manner is a useful tool to study the mechanochemical transformations that accompany cellular processes and to obtain insights about the mechanisms involved. By monitoring the response to an applied force of a molecule undergoing one of these processes, it is possible to determine the thermodynamic and kinetic parameters of the reaction. However, it was the invention of optical tweezers by Ashkin that gave researchers the ability to investigate these force-generating cellular processes [3].

The principle of optical tweezers is based on the radiation pressure exerted by light when colliding with matter. The effect of light on matter was first speculated by the 17th-century astronomer Johannes Kepler, who observed that comet tails always point away from the sun [4]. Lebedev was the first to measure experimentally the radiation pressure posited by Maxwell–Bartoli, and showed that the pressure for a reflective surface is twice that of an absorbing surface [5]. The advent of lasers in the 1960s finally enabled researchers to study radiation pressure through the use of intense, collimated sources of light [6, 7]. In 1970, Ashkin observed that micron-sized latex spheres, or beads are attracted toward the center of an argon laser beam with the power of a few mW [8]. By using two counter-propagating beams, he created the first stable optical trap for beads suspended in water. It was not until 1986 that Ashkin, together with Chu and others demonstrated the present form of optical tweezers that uses a single, tightly focused laser beam to stably trap 25 nm to 10 mm particles in three dimensions [3].

Optical tweezers were immediately found to be useful in biological research, due to their ability to trap and move microorganisms without physical contact, which can even allow manipulation of organelles in live cells [9]. In addition, the ability to measure small forces, from femtonewtons to some tens of piconewtons, has made optical tweezers a star player in quantitative biophysics and mechanobiology [10]. In this chapter, our aim is to provide a comprehensive overview of optical trapping theory, and considerations for instrument design and calibration which helps biophysicists, biochemists, and cell biologists to build and calibrate their own instruments and to perform single-molecule force measurements on both in vitro systems and in living cells. Finally, we will attempt to communicate our sense of future directions for this growing field.

2.2 Optical Trapping Theory

The origin of optical forces is from the exchange of momentum between the incident photons and an optically trapped object [11]. The forces exerted by photons on the object can be divided into two components: the scattering force which pushes the object away from the light source and the gradient force which pulls the object toward the region of highest light intensity, i.e., the focal region of the light (Figure 2.1) [3]. The scattering force can be regarded as a consequence of the momentum delivered by the scattered photons, while the gradient force arises from the interaction between the light-induced dipoles and the inhomogeneous electric field in the direction of the field gradient. Although the full theory of optical trapping is quite complex, some informative approximations are available in some extreme conditions.

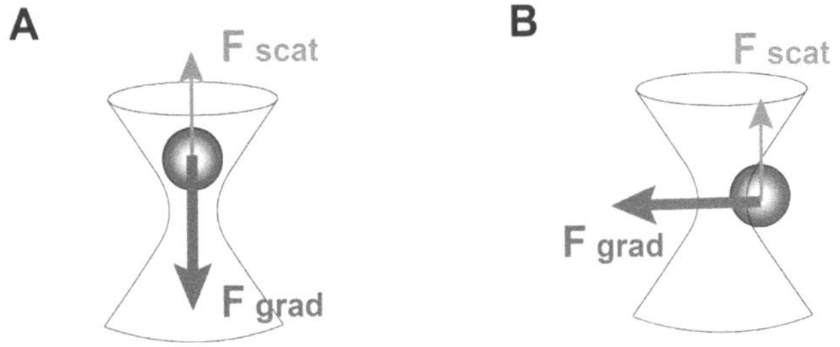

Figure 2.1 Schematic of the optical forces on an optically trapped particle. The axial (A) and radial (B) gradient force is the dominant component to form an optical trap.

2.2.1 Rayleigh Optics Approximation

When the trapped sphere is much smaller than the wavelength of the trapping laser, i.e., $r << \lambda$, the conditions for Raleigh scattering are satisfied and the particle can be treated as a collection of dipoles polarized by the envelope of the light field forming the trap, with the phase of the field being approximately constant throughout the particle [12]. The scattering force is due to the absorption and re-radiation of light by the dipole. For a spherical particle of radius r, this force is

$$F_{\text{scat}} = \frac{I_0 \sigma n_c}{c} \tag{2.1}$$

Where σ is the optical cross section of the particle, $n_c = n_p/n_m$ is the ratio of the indices of refraction of the particle (n_p) and of the medium (n_m) [13, 14]. If $r \ll \lambda$, the Rayleigh scattering cross-section is:

$$\sigma = \frac{8}{3}\pi (kr)^4 r^2 \left(\frac{n_c^2 - 1}{n_c^2 + 2} \right)^2 \tag{2.2}$$

where $k = 2\pi n_m/\lambda$ is the wave number of the light. The scattering force on a Rayleigh particle can then be written in terms of the light intensity I_0:

$$F_{\text{scat}} = \frac{128\pi^5 r^6}{3\lambda^4} \left(\frac{n_c^2 - 1}{n_c^2 + 2} \right) \frac{n_m}{c} I_0 \tag{2.3}$$

Thus, the scattering force is dependent on the light intensity, the wavelength of the trapping light, the particle size, and its refractive index contrast against the liquid in which it is immersed.

The gradient force can be calculated from the Lorentz force on the induced dipole by the light field. The induced dipole moment of the dielectric sphere can be written as

$$\vec{p} = \alpha \cdot \vec{E} \tag{2.4}$$

where $\alpha = 4\pi n_m^2 r^3 \left(\frac{n_c^2 - 1}{n_c^2 + 2} \right)$ and is the relative polarizability of the particle. The interaction between the induced dipole and the electric field of the light (E) creates an electrostatic potential:

$$U = -\vec{p} \cdot \vec{E} \tag{2.5}$$

Thus the gradient force is

$$\vec{F}_{\text{grad}} = -\nabla U = -p\nabla\vec{E} = -\alpha\left(E \cdot \nabla\right)\vec{E} = \frac{2\pi n_m}{c}\left(\frac{n_c^2 - 1}{n_c^2 + 2}\right)r^3\vec{\nabla}I \tag{2.6}$$

For a beam with a Gaussian profile, the intensity gradient is approximately linear in the vicinity of the center of the beam.

2.2.2 Ray Optics Approximation

When the trapped sphere is much larger than the wavelength of the trapping laser, i.e., $r >> \lambda$, the optical forces can be calculated from simple ray, or geometrical optics [15]. Refraction of the incident light by the sphere corresponds to a change in the momentum carried by the light. The basic geometry for ray-optic calculations can be seen in Figure 2.1. The force on the sphere, thus, can be directly computed by the rate of momentum change of the light:

$$F_{\text{scat}} = \frac{n_m \cdot P}{c}\left(1 + R\cos\left(2\theta_R\right) - \frac{T_F^2\left(\cos\left(2\theta_R - 2\theta_T\right) + R\cos\left(2\theta_R\right)\right)}{1 + R^2 + 2R\cos\left(2\theta_T\right)}\right) \tag{2.7}$$

$$F_{\text{grad}} = \frac{n_m \cdot P}{c}\left(R\sin\left(2\theta_R\right) - \frac{T_F^2\left(\sin\left(2\theta_R - 2\theta_T\right) + R\sin\left(2\theta_R\right)\right)}{1 + R^2 + 2R\cos\left(2\theta_T\right)}\right) \tag{2.8}$$

where θ_R is the angle of incidence, θ_T is the angle of refraction, R and T_F are the Fresnel reflection and refraction coefficients.

2.2.3 Electromagnetic Theory (MDSA Approximation)

In the typical application of optical tweezers in biological research, the size of trapping particles is usually comparable to the wavelength of the trapping laser, i.e. $r \sim \lambda$, and neither Rayleigh nor Ray optics approximation are appropriate. For such particles, diffraction effects are significant, and the vector character of the electromagnetic field cannot be neglected, as is generally the case in scalar theories that use a paraxial approximation. These factors increase the difficulty enormously of computing realistic forces in this intermediate size regime. To address these questions, Nussenzveig et al. derived the Mie–Debye-spherical aberration (MDSA) theory, named after Mie and Debye to calculate the trapping force generated by optical tweezers, and this is currently the most complete and realistic theory of optical tweezers [5, 13, 16–18].

In this theory, the scattering force is calculated based on classical diffraction theory, which is called the Lorenz–Mie–Debye solution. The highly focused laser beam is Mie scattered by the particle, and the force F_{scat} can be obtained by computing the Maxwell stress tensor for the total (incident and scattered) field and integrating over the surface of the microsphere. It is usually expressed in terms of an axial, dimensionless efficiency factor Q_z, the ratio of F to one half the force $2P/\upsilon$ exerted on a perfectly reflecting mirror by a perpendicularly incident light beam of power P in a medium of refractive index n_m, where $\upsilon = c/n_m$,

$$Q_z = \frac{cF}{n_m P} \tag{2.9}$$

At an axial equilibrium position z_{eq}, the axial trap stiffness k_{scat} is given by

$$k_{\text{scat}} = -\frac{n_m P}{c}\left(\frac{\partial Q_z}{\partial z}\right)_{z=z_{\text{eq}}} \tag{2.10}$$

In the geometrical optics limit, Q_z depends only on the dimensionless parameter $\zeta \equiv z/r$, so that

$$\frac{\partial Q_z}{\partial z} = \frac{1}{r}\frac{dQ}{d\zeta} \tag{2.11}$$

In this limit, the scattering force stiffness must therefore decay hyperbolically as a function of the microsphere size r, providing an important validation test. Using an MDSA approximation, the gradient trap stiffness k_{grad} could be obtained:

$$k_{\text{grad}} = -\frac{n_m P}{c}\left(\frac{\partial Q_\rho}{\partial \rho}\right)_{z=z_{\text{eq}}} \tag{2.12}$$

where Q_ρ is the transverse dimensionless efficiency factor.

2.3 Optical Tweezers Instrumentation

The optical tweezers can be designed in various geometries based on particular research needs [11], but they all share common components: (i) the laser and power stage, which controls the power of the trap; (ii) beam steering components, which manipulate and move the trap; (iii) focusing optics, which usually consist of a microscope objective with a high numerical aperture (NA); and (iv) the position detection components. Figures 2.2 and 2.3 show two typical optical design of instrument developed in the Bustamente Lab at University of California, Berkeley [19, 20]. We next discuss the design principles and details of the choice of each component of two instruments.

2.3.1 Optical Setup

Figure 2.2 shows the design of a dual-beam optical tweezers instrument which is also called "MiniTweezers" [19]. In this instrument, two counter-propagating laser beams are focused on the same spot by two identical water immersion microscope objectives. The trapping laser itself can be used to measure the bead position. The two laser beams are moved and shaped synchronously by two piezo-based fiber-wigglers with a closed-loop control system. The forces exerted on the bead could be calculated by measuring the change in the momentum flux of the light beam as it enters and leaves the trap measured by four position-sensitive photodetectors (PSDs). The major advantages of this design are its high integration level, low cost, and low sensitivity to environmental vibration.

Figure 2.3 shows the design of a high-resolution dual-trap optical tweezers [21–26]. In this instrument, a linearly polarized laser beam is split by polarization at the first polarizing beam splitter (PB1). The s-polarized beam is deflected by a two-axis piezosteerable mirror (SM), while the p-polarized beam is deflected by a fixed mirror, and the two beams are then recombined at the second polarizing beam splitter (PB2). A second telescope (T2) provides the final magnification of the beams, and images the plane of the steerable mirror onto the back focal plane of a 1.2 NA 60× water immersion objective (O1) (Nikon, Melville, NY). The objective focuses the beams on two diffraction-limited spots in the center of a 200 μm-thick fluidics chamber. A second, identical objective (O2) collects the forward scattered light, which is split again by polarization at a third polarizing beam splitter (PB3) and imaged onto two separate PSDs (PD1 and PD2). A light-emitting diode provides light for Kohler illumination, and the second objective and an additional tube lens image the specimen plane onto a charge-coupled device camera (CM).

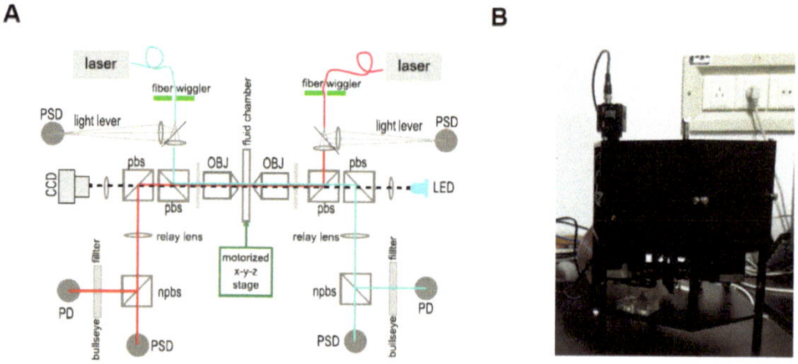

Figure 2.2 (A) Schematic layout of MiniTweezers, designed by Smith et al. [19]. Diode lasers (LU0808M250, 808 nm, 250 mW) are mounted on temperature-controlled mounts, beam circularized with anamorphic prisms, and protected from reflections by Faraday-effect opto-isolators. A spatial filter (two 100 mm lenses with a 40 μm pinhole) passes ~80% of the laser power. Polarizing beam splitters (PBS) separate different polarizations for infra-red beams but pass blue light in either polarization for a charge-coupled device (CCD) camera image. Quarter waveplates (QWP) (CVIQWPO-838-05-4) are used for circularly polarized beam at foci. (B) Photograph of the MiniTweezers instrument in our lab.

2.3.2 Trapping Laser

To obtain a stable trap, the trapping laser should have a high pointing stability and low power fluctuations. Furthermore, a Gaussian mode is focused on the smallest diameter beam waist and will therefore produce the most efficient, harmonic trap. In addition, the wavelength of the laser is also an important consideration when biological material is to be trapped. In practice, the typical lasers used for optical tweezers are single-mode lasers in the near infrared (800 nm–1100 nm) owing to the high beam powers that are available and the minimal damage to cells at these wavelengths [27].

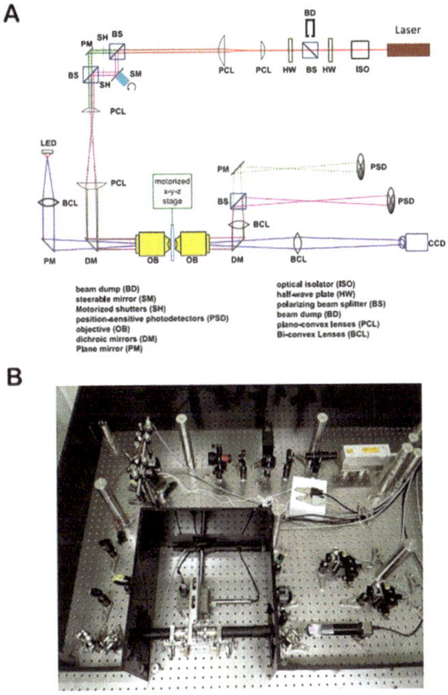

Figure 2.3 (A) Schematic layout of the dual-trap optical tweezers, designed by Bustamante et al. [25]. (B) Photograph of high-resolution dual-trap instruments in our lab.

2.3.3 Beam Steering Module

The ability to manipulate and move the object inside the trap, relative to the position of the sample stage (surface or pipette-sucked beads) is a fundamental element of the optical tweezers. Beam steering elements should be placed at optical planes conjugate to the back-focal plane of the objective, so that beam rotations become translations in the sample plane [28]. Several sensitive optical methods have been used to control the position of a trap and these include closed-loop piezo-driven mirrors, acousto-optic deflectors (AODs) or electro-optic deflectors [11]. Specifically, using a closed-loop piezo-driven mirror to reflect the incident beam is the easiest way to precisely control and move the trap. The comparatively slow temporal response limits their usefulness for fast-scanning applications, but their low insertion loss and large deflection angles make them an inexpensive option for slow-scanning and feedback applications [25].

On the other side, an AOD contains a transparent crystal inside of which an optical diffraction grating is generated by the density changes associated with an acoustic traveling ultrasound wave. By altering the acoustic frequency to the deflector, the angle through which the diffracted beam is deviated can be varied. The angular range of an AOD is typically much larger than that of a piezo mirror. In addition, the angular modulation rate is very high, reaching ~10 kHz, as opposed to hundreds of Hz for piezo-driven mirrors. However, due to the optical loss in the AODs, its transmission efficiency is typically only ~60%, and this efficiency can vary by a few percent as a function of deflection angle, leading to somewhat significant trap stiffness variations. Therefore, both piezo-driven mirrors and AODs have their advantages and disadvantages, and the choice between them depends on the relative importance of the manipulation speed, power loss, and trap stiffness variation.

2.3.4 Trapping Objectives

Optical tweezers are often built on a conventional light microscope, which provides a ready-made platform for focusing the laser beams and imaging the samples. The objective is the most important single element in an optical tweezers instrument, and the choice of an appropriate objective determines the overall efficiency of the optical tweezers. In principle, the higher NA the objective has, the higher gradient force the optical tweezers can reach. A high NA objective (typically, 1.2–1.4 NA) is required to produce an intensity gradient sufficient to overcome the scattering force and produce a stable optical trap for microscopic objects, such as polystyrene beads. Oil immersion objectives typically have higher NAs than their water immersion counterparts. However, the working distance of most high NA oil immersion objectives is quite short, and the large refractive index mismatch between the immersion oil and the aqueous trapping medium leads to significant spherical aberrations [29]. Furthermore, the drag coefficient of a bead trapped near a surface is increased due to hydrodynamic coupling with the surface. Therefore, the water immersion with moderate NA (typically 1.2) would be a proper choice for trapping deeper objects in solution [25].

2.3.5 Position Detection

The key to quantitative optical trapping is accurate detection of the position of the particle in an optical trap. Within the volume of the laser focus, the displacement of the particle from its equilibrium position is directly proportional to the forces acting on this particle. Many methods could be used for the detection of these displacements, and the two most commonly applied methods are video-based imaging, and back-focal-plane interferometry (BFPI)-based position detection [16]. The simplest method is to image the particle directly using a video camera. The position of the trapped object is determined and tracked through specific video tracking algorithms [30]. However, the spatial and temporal resolution is ultimately limited by the imaging speed of the camera and the memory capacity of the computer.

Another common technique used in optical tweezers systems is the BFPI-based assay. Since first proposed by Schmidt et al. in 1998 [31], the BFPI assay has found increasing acceptance for nanometer precision measurements of the displacement of trapped samples. In this assay, light passing through the sample is collected by a PSD that is optically conjugated to the back-focal-plane of the condenser lens. This detection is largely independent of the position of the trap in the field of view, and as a result, the detector response is insensitive to the location of the trap. The interference pattern created at the objective back focal plane between the light forward-scattered by a trapped bead and the transmitted light can be sensitively detected by the PSD or a quadrant photodiode detector (QPD). According to the approximate model by Schmidt et al., based on Rayleigh scattering, the differential output signal of PSD with a small lateral displacement x is expressed below:

$$\frac{I_+ - I_-}{I_+ + I_-} \approx C\frac{kd^3x}{w_0^3}e^{-\left(\frac{x}{w_0}\right)^2} \tag{2.13}$$

Where $\frac{I_+ - I_-}{I_+ + I_-}$ is the differential output of the QPD or the PSD, and corresponds directly to the position of the bead; C is the correction coefficient, d is the diameter of the bead, k is wave vector of the detecting light, x is the lateral displacement of the bead with respect to the trap center, w_0 is the waist size ($1/e$ radius of the focus).

For $x << w_0$, the response is linearly proportional to x, indicating a linear relationship between the relative position of the bead with respect to the trap center and the PSD response.

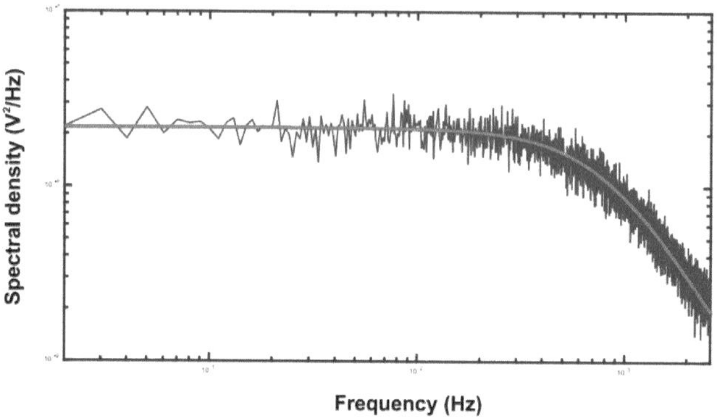

Figure 2.4 Power spectrum of a trapped bead. The raw power spectrum (blue line) was fit (red line) to Eq. (2.14).

BFPI typically magnifies lateral deflections of the bead several thousand-fold, which is much higher than that in imaging based methods.

2.4 Force Calibration in Optical Tweezers

For small motions of a bead near the center of an optical trap, the force (F) exerted on the bead is proportional to the displacement of the bead from the trap center (x): $F = -k \cdot x$, where k is trap stiffness. In order to determine the force directly, the trap stiffness of the instrument must be calibrated. Considerable effort has been made to measure the trap stiffness with high accuracy. Many methods have been developed for this purpose, three of which will be described in detail here.

2.4.1 Viscous Drag Force Calibration

The easiest way to calibrate the trap stiffness is to measure the displacement of the trapped bead when subjected to a viscous drag force produced by moving the bead through fluid [32]. If a liquid with viscosity η flows past a bead of radius r with velocity v, the force

(F) due to viscous drag could be described by $F = 6\pi\eta rv$. Since the viscosity of the liquid is known and beads of known radius can be obtained, we can apply a known force provided that the velocity of the liquid is known. The trap stiffness could then be determined by measuring the displacement of the bead (x) by the viscous drag. In addition to its simplicity, another advantage of this method is that it could calibrate the region over which the trap is harmonic.

2.4.2 Brownian Motion Calibration

Another common calibration method relies on the thermal motion of a bead of known size in the optical trap [33, 34]. The thermal motion of a spherical bead of known size suspended in water has been well characterized. As the bead is trapped by the optical tweezers, its Brownian motion is constrained to the center of the trap. The one-sided power spectrum for a trapped bead is:

$$S_{xx}(f) = \frac{k_B T}{\pi^2 \beta \left(f_0^2 + f^2\right)} \tag{2.14}$$

where $S_{xx}(f)$ is in units of displacement/Hz and can be calculated from the PSD signal, k_B is Boltzmann's constant, T is the absolute temperature, β is the dynamic drag coefficient of the bead, f_0 is the rolloff frequency, which is related to the trap stiffness k: $f_0 = k/2\pi\beta$. For a sphere bead of radius r in a medium with viscosity η, the drag coefficient β is given by usual Stokes relation: $\beta = 6\pi\eta r$. Therefore, the trap stiffness could be obtained through fitting the power spectrum with an overall scaling factor to the rolloff frequency f_0 (Figure 2.4).

2.4.3 Direct Measurement

It is known that optical forces originate from the exchange of momentum between the incident photons and the optically trapped object [19]. The value of the force can be obtained by integrating the light intensity entering (S_{in}) and leaving (S_{out}) the trapped bead:

$$F = \left(\frac{n_m}{c}\right) \iint \left(S_{\text{in}} - S_{\text{out}}\right) dA \tag{2.15}$$

where n_m is the refractive index of the medium and c is the speed of light. For counter propagating beam rays used in MiniTweezers systems, it is possible to calculate the momentum transfer directly by collecting all of the light scattered by the trapped particle by the PSD. The intensity pattern $I(x,y)$, collected by the PSD, produces an intensity signal given by:

$$S_x = \psi \iint \frac{x}{R_D} I(x, y) dxdy \tag{2.16}$$

where R_D and ψ are the size and the efficiency of the detector, respectively. The gradient force of the trap F_x and F_y can then be obtained in terms of the detector signals and known constants:

$$F_x = \frac{D_x R_D}{c\psi R_L}, \quad F_y = \frac{D_y R_D}{c\psi R_L} \tag{2.17}$$

where D_x and D_y are the x and y signal of the PSD and R_L is the focal length of the lens. In principle, this approach is applicable to any optical trapping configuration. However, because it necessitates measurement of the total intensity of scattered light, it has only been implemented for relatively low NA, counter propagating optical traps, where the microscope objective entrance pupils are underfilled.

2.5 Combined Optical Trapping and Single-Molecule Fluorescence Spectroscopy

Optical tweezers are powerful tools for studying single biological molecules, but there are many properties that cannot be probed simply by using traditional optical tweezers. The single-molecule force- and fluorescence-based approaches are highly complementary: force spectroscopy reports on more global structural and mechanical rearrangements in biomolecules, while fluorescence measures local conformational changes. Therefore, hybrid instruments combining fluorescent capabilities and mechanical manipulation provide a promising direction for single-molecule spectroscopy, and recently some instruments have been successfully constructed and have been used to study the motion in myosin, the dynamics of DNA and nucleosome, and the DNA-protein interactions [35–42].

The main challenge for combining optical traps with fluorescence microscopy is the photobleaching of the fluorophores near the optical trap. Therefore, simultaneously exposing the fluorescence labeled the sample to the trapping and excitation lasers should be avoided. One simple solution is to separate the optical trap and fluorescence-labeled molecule by using a long DNA molecule to tether the system of interest to the trapped bead. By using a 15 μm DNA molecule as the tether, Hohng et al. successfully detected the conformational dynamics of Holliday junction under different forces [37]. However, the long DNA molecules are very flexible and thus poor mechanical motion transducer, and decrease the resolution spatial resolution of the whole system. Another solution is to separate the trapping and fluorescence excitation temporally, i.e., turning them on and off out of phase [36]. By using this approach, Comstock et al. build the first high-resolution optical tweezers with single-molecule fluorescence detection, and enable the simultaneous measurement of angstrom-scale mechanical motion of individual DNA-binding proteins along with the detection of properties of fluorescently labeled protein [38].

2.6 Nanophotonic Optical Tweezers

The traditional optical tweezers typically use a microscope objective to form an optical trap and manipulate particles at the microscale. Their fundamental drawback is the attainable resolution, which is typically on the order of the wavelength of light given by the diffraction limit $d_{\text{min}} = 1.2\lambda/\text{NA}$, where NA is the numerical aperture, which makes it difficult to trap objects at the nanoscale. Although several different strategies to trap nanoscale particles have been proposed, such as use of oil-immersion optical lenses with high NA value, they need however complicated tabletop equipment and are quite bulky [43, 44]. Recently, the development of near-field optical microscopy offers an alternative strategy to achieve a large optical trapping force [45]. Near-field devices provide a higher electromagnetic energy gradient, because the field exponentially decays from the surface where the light is confined. The rapid exponential decay generates a strong trapping force and enables nanoscale manipulation. In 2008, Grigorenko et al. first reported plasmonic optical traps in the proximity of closely spaced pairs of gold nanodots [46]. The near-fields of the nanodot pairs produced subwavelength trapping volumes where nanometer-sized polystyrene beads could be stably captured. This opened up exciting new possibilities for the nanomanipulation of biological samples with plasmonic trapping. After the pioneering work of Grigorenko, various plasmonic nanostructures with different morphologies, such as double-nanoholes, nanoapertures, and nanodipole antennas, were employed to trap small particles, or even *Escherichia coli* bacteria [47–51]. Figure 2.5 shows the three-dimensional optical manipulation of single 50 nm polystyrene bead by a nano-optical trap built by engineering a bowtie plasmonic aperture at the extremity of a tapered metal-coated optical fiber [51]. Both the trapping operation and monitoring are performed through the optical fiber, making it totally autonomous and free of bulky optical elements. The achieved trapping performances allow for the trapped specimen to be moved over tens of micrometers over a period of several minutes with very low in-trap intensities.

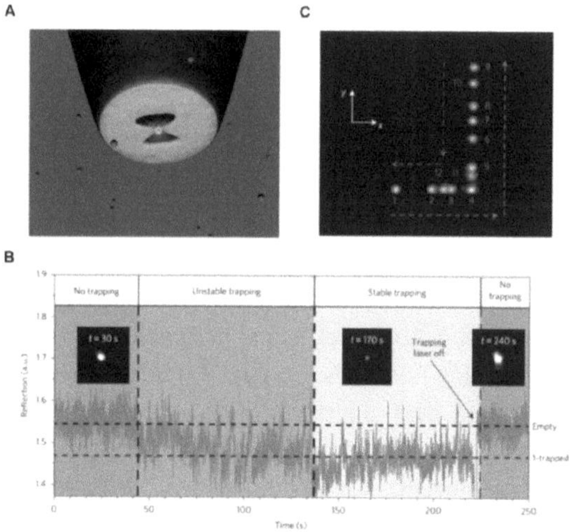

Figure 2.5 Three-dimensional manipulation of a single 50 nm polystyrene bead by the nanophotonic optical tweezers. (A) A 1,064 nm trapping laser is directly coupled into the metal-coated tapered optical fiber patterned with a bowtie nano-aperture. The plasmonic trap is generated at the extremity of the tapered optical fiber. (B) Time trace of the reflected 1,064 nm signal showing trapping of a single 50 nm polystyrene bead. (C) Composite image reproducing the displacement of the trapped object. Numbers 1–12 represent the successive steps of the tip movement. Reprinted with permission from ref. [51].

2.7 Applications of Optical Tweezers in Single-Molecule Manipulation

Optical trapping allows one to manipulate individual molecules, as well as to measure the forces exerted during various biological processes. The practical use of optical tweezers in single-molecule biophysics has been summarized in many excellent reviews [10, 52–56]. Here, we will choose some specific examples to highlight some of the useful insights provided by optical tweezers into important biological processes.

2.7.1 Mechanical Properties of DNA

The mechanical properties of the DNA double helix are unlike those of any other natural or synthetic polymer. As a model polymer, its elastic properties have been probed extensively by optical trapping measurements and direct measurements of force and extension on single molecules of DNA have provided the most rigorous test to date of theories of entropic elasticity [52, 57–59]. As seen in Figure 2.6, the force–extension curve (FEC), which shows a characteristic relationship between the force (F) on the molecule and the extension of the molecule (x), could be closely fitted with a worm-like chain model. An interpolation formula for worm-like chains have been used to fit FECs of dsDNA :

$$F(x)=\frac{k_B T}{L_p}\left[\frac{1}{4}\left(1-\frac{x}{L_c}+\frac{F}{K}\right)^{-2}-\frac{1}{4}+\frac{x}{L_c}-\frac{F}{K}\right] \tag{2.18}$$

where L_p is the persistence length, L_c the contour length, and K the elastic modulus of the dsDNA. The persistence length of dsDNA is typically ~50 nm, and the elastic modulus is ~1,000 pN to 1,200 pN [59]. Because L_p and L_c are well known for dsDNA, it is usually used to as a "handle" for attaching a molecule of interest to the beads.

Biochemical and biophysical studies showed that the helical structure of DNA is highly adaptable and can assume various forms. Single-molecule manipulation experiments have revealed the existence of an additional helical form of DNA stabilized by external forces. When tension in a nicked DNA molecule is increased to around 65 pN, it displays a reversible, cooperative transition, an overstretching transition to an extended form that is ~70% longer than classical right-handed helical form of DNA (B-DNA) and with a substantially reduced twist. The physics underlying dsDNA overstretching transition is complex, and three mechanisms have been proposed: strand unpeeling, localized base-pair breaking (melting), and formation of S-DNA (stretched form DNA). Recent studies indicated that all these three transitions can exist, and the balance between the three structures of overstretched DNA is governed by both DNA topology and local DNA stability [60].

The mechanical properties of single-stranded DNA (ssDNA) were also characterized using optical tweezers. It was found that ssDNA is more resistant to force at low forces but more compliant at high forces. The freely jointed chain model can be used to fit to the FECs of ssDNA, producing a persistence length of 0.75 nm and elastic modulus of 800 pN [57].

2.7.2 Folding and Structural Dynamics of Proteins and Nucleic Acids

How bio-macromolecules, such as proteins or nucleic acids, fold into their specific three-dimensional structures is one of the central topics in biology because of the close relationship between structure and function. Traditional methods to study the folding of these molecules are based mainly on monitoring the structural changes using various forms of spectroscopy such as circular dichroism, nuclear magnetic resonance spectroscopy, and fluorescence spectroscopy during the thermal or chemical denaturation. However, all these techniques provide ensemble averaged signals from a large number of molecules, which restricts the information from rare conformers in the solution. When a very small force by optical trap is applied to the two ends of a linear biopolymer, such as proteins and nucleic acids, the molecule can be denatured by the mechanical tension, and the structural change of the molecule can be monitored by its FECs. The ability to measure individual folding events by single-molecule manipulation allows direct observation of rare or transient folding intermediates. Moreover, if the process is reversible, it is possible to extract both kinetic and thermodynamic information from these experiments at the same time as the forces that maintain the three-dimensional structure of the molecule in solution are characterized [61]. Therefore, single-molecule force spectroscopy is an ideal method to study the folding of biomolecules. Experimentally, two forms of measurements are commonly used: nonequilibrium force-ramp and equilibrium constant-force measurements.

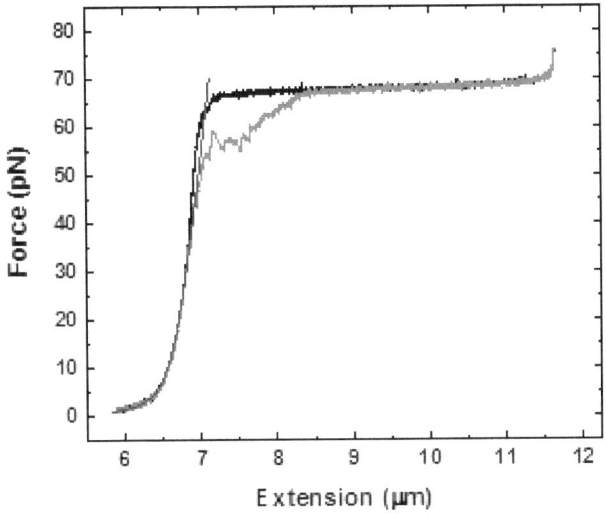

Figure 2.6 FEC of double-stranded DNA (dsDNA) using optical tweezers. The black line shows the stretching of dsDNA, while the red line shows the relaxation of the stretched DNA. The blue line shows the worm-like chain model (Eq. (2.18)) fitting of the stretching trace.

The first type of measurement is to record the extension changes of the molecule accompanied by changes in force (force-ramp measurements). This produces a characteristic sawtooth pattern of rips for each structure that unfolds (Figure 2.7A). Both the unfolding force and the extension change are important to reconstruct the folding energy landscape of the molecule. The extension change could be used to determine precisely the structural features of the molecules. The distribution of unfolding force could be fitted by the Dudko model to obtain important kinetic information [62–66], such as unfolding rate at zero force k_{off}, the height of the energy barrier $\Delta G^{\ddagger}$ and the distance to the transition state $\Delta x^{\ddagger}$. In addition, thermodynamic equilibrium-free energies could also be recovered from these nonequilibrium data by using recently published fluctuation theorems, such as Jarzynski's equality [67]. In fact, the first experimental demonstration of the validity of fluctuation theorems was obtained by mechanically unfolding a single molecule of RNA with optical tweezers [68].

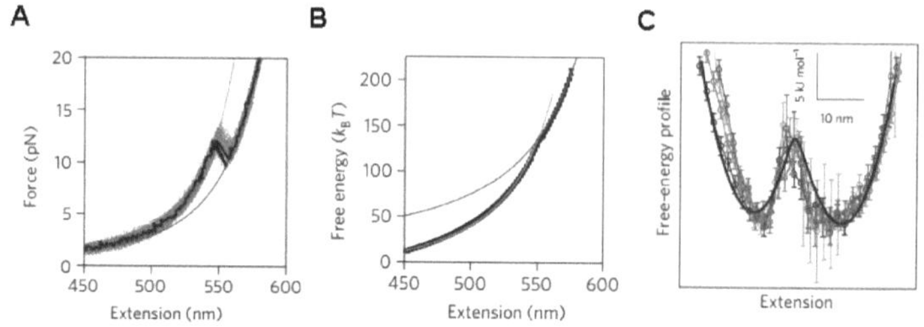

Figure 2.7 Reconstruction of the energy landscape from force-ramp measurements. (A) Multiple FECs of a DNA hairpin construct. (B) The free-energy landscape at zero force reconstructed from the force ramps is dominated by the stretching energy of the DNA handles. (C) The landscape from the force ramps (red), tilted to 14 pN, agrees well with the profile from a constant-force measurement of the same hairpin (black) before deconvolution. Reprinted with permission from ref. [65].

The second type of measurement records the extension changes of the molecule as a function of time while the applied force is kept constant. When the force reaches near the equilibrium value, the folding and unfolding states of the molecule keep switching with each other, resulting in fluctuation of the extension with time. The extension change for each folding transition can be measured as the distance between peaks in histogram of the records. The distribution of lifetimes of the states can also be directly determined. The force dependence of the rates is particularly informative, since the kinetics can be extrapolated to zero force, and the variation of the lifetimes with force yields the location of the transition states, and the variation of the lifetimes with force yields the location of the transition state. These interpretations are, however, model dependent. The simplest interpretation, originally proposed by Bell [69], assumes that the transition-state location does not depend on force, and takes the rates for a two-state reaction as:

$$k_u(F) = k_{0,u}\exp\left(\frac{\Delta x_u^{\ddagger}}{k_B T}F\right) \tag{2.19}$$

where $k_{0,u}$ and $k_{0,f}$ are the rates for unfolding and folding at zero force, respectively, and $\Delta x^{\ddagger}_{u}$ and $\Delta x^{\ddagger}_{f}$ are the distances to the transition state from the unfolded and folded states, respectively. Constant-force extension measurements allow a direct determination of the shape of the folding energy landscape along the entire reaction coordinate. The simplest method for reconstructing energy landscape profiles is based on obtaining the probability distribution $P(x)$ of the molecular extension x, from equilibrium constant-force measurements. The desired landscape profile is the equilibrium-free energy function $G(x)$, which could be found directly from an inverse Boltzmann relation:

$$G(x) = -k_B T \, ln\left[P(x)\right] \tag{2.20}$$

where $k_B T$ is the thermal energy [70]. This approach has been widely used to study the folding of nucleic acids and proteins [71–75].

2.7.3 Dynamics of Molecular Motors

Molecular motors are biological molecular machines which transform chemical energy into motion and mechanical work, and are essential in many biochemical and biophysical processes, including gene replication, transcription, translation, cellular transport, cell motility and cell division [76, 77]. Understanding the molecular details of the working mechanism of these motors is one of the most important fundamental scientific problems. By directly observing the motion of individual motor proteins, an optical tweezers setup has been widely used to study the dynamics of molecular motors, and has provided much important information on how molecular motors operate [53, 56, 78]. Cytoskeleton bound motor proteins, such as dynein, kinesin, and myosin, utilize the energy from hydrolysis of adenosine triphosphate (ATP) or related compounds to move and transport cargo along the cytoskeleton filaments (actin filaments or microtubules). Using optical tweezers, Block et al. systematically studied the motion of single molecules of kinesin. They first observed the hand-over-hand walk of kinesin along microtubules, and further found that these walks are uneven, with an alternation in the dwell times of successive steps [79–81]. Spudich et al. subsequently observed similar steps for myosin [82, 83]. In contrast to the relatively strong interaction between kinesin and microtubule, dynein binds weakly to microtubules. However, using optical tweezers, Mallik et al. found that groups of dynein motors work well together to generate large forces that tune linearly in strength and persistence with dynein number [84]. By measuring velocity as a function of force and ATP concentration, it is possible to determine the coupling of the motor's biochemical and mechanical cycles, and obtain the kinetic model for these motors. Mechanical steps could be modeled as a transition over a free energy barrier, where the time to transition rises exponentially with the applied force following an Arrhenius–Boltzmann relation [85]:

$$k(F) = k_u + k_f \exp\left(-\frac{F\delta}{k_B T}\right) \tag{2.21}$$

in which k_u includes the rates for all force-independent transitions, k_f is the scaling prefactor for the force-dependent rates, and δ is the distance along the reaction coordinate to the mechanical transition state. At zero force, the unloaded rate $k_0 = k_u + k_f$.

Optical tweezers have also been used to probe the motion and mechanisms of nucleic acids motors such as the DNA packaging motor, RNA polymerase, and ribosome. The machinery involved in the packaging of viral DNA has two components, the portal-connector and ATPase. Through systematic studies of the packaging of viral DNA into the phi29 bacteriophage, Bustamante et al. demonstrated that the rotary portal motor of the bacteriophage can package DNA against high forces, and elucidated a minimal kinetic model of force generation [86–90]. RNA polymerase is the enzyme responsible for copying the information stored in a DNA sequence into the messenger RNA during the process of transcription. Using the high-resolution optical tweezers with single base-pair resolution, the groups of Bustamante and Block systematically studied the individual transcription events, characterized their heterogeneity, revealed their stochastic alternation in periods of continuous translocation and pauses, and provided kinetics models of chemo-mechanical coupling in transcription [91–101]. Ribosomes are complex molecular machines that hydrolyze GTP to translate the information encoded in mRNA into proteins. During protein biosynthesis, the ribosome moves along the mRNA in the 5′-to-3′ direction, catalyzed by the forward translocase elongation factor G (EF-G). Through subtle design, the group led by Tinoco and Bustamante successfully used optical tweezers to study translation by a single ribosome on one mRNA (Figure 2.8) [102–110]. They showed that translation of a single messenger RNA by a ribosome occurs by successive translocation-and-pause cycles. They further studied the modulation of nascent protein folding by the ribosomal environment, and provided a kinetic model describing how a protein can regulate its own synthesis by the force generated during folding [107]. Through direct measurement of ribosome-dependent mRNA dynamics during programmed frameshifting, they demonstrated multiple ribosomal translocation attempts while in register with a slippery sequence [108].

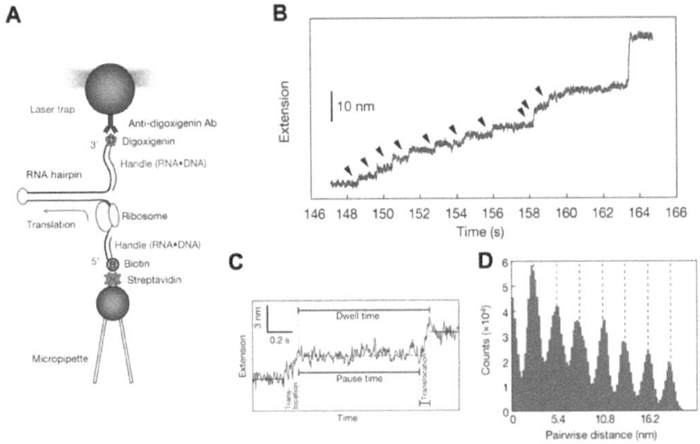

Figure 2.8 Single-molecule translation using optical tweezers. (A) Experimental design. The ribosome was stalled at the 5′ side of the mRNA hairpin construct. (B) Extension and force trajectories during translation. Discrete steps are indicated by arrowheads. (C) The pattern of the translation process illustrated by a close-up view of the 148 s–150 s region. (D) Pairwise distance analysis of the extension trajectory in from 47 s to 157 s after correction for drift. Reprinted with permission from ref. [102].

2.8 Summary and Perspective

Over the 30 years of development, optical tweezers-based single-molecule manipulation has been applied to study a wide range of biophysical problems. This new technique has yielded important insights into RNA and protein folding, motion, and mechanisms of biological machines such as nucleic acids motors and cytoskeleton filaments motor proteins. However, although the progress to date has been extraordinary, there are still many challenges and limitations yet to be overcome in the future. Improvements in instrumentation and biochemical sample preparation will lower the barriers for studying highly complex biological processes with optical tweezers, and the combination of mechanical manipulation and fluorescence detection of single molecules will further expand the range of questions that can be addressed by adding an independent reaction coordinate.

Appendix: A Tutorial Protocol: Single-Molecule Mechanical Folding of Hairpin DNA Using Optical Tweezers

Purpose

The ability to obtain a nanometer spatial resolution, a piconewton force resolution, and a millisecond time resolution by optical tweezers makes it possible to study folding of proteins or nucleic acids and enables monitoring of the folding events and fluctuations between different molecular conformations. Here, we provide a detailed protocol to study folding of DNA hairpin using optical tweezers (Figure 2.9A), including sample preparation, data collection, and methods of data analysis to extract folding energies and rates from the single-molecule measurements.

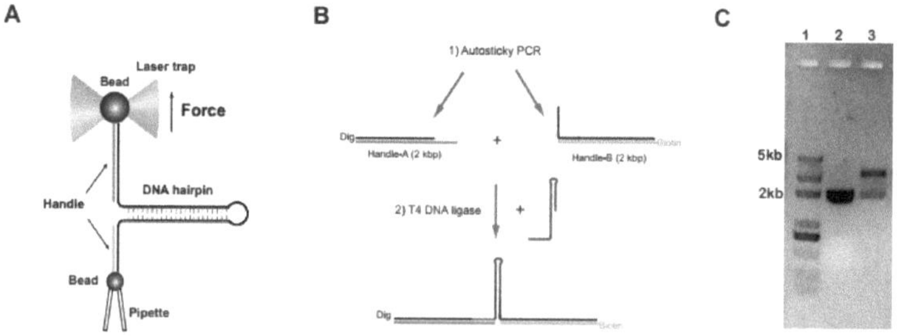

Figure 2.9 (A) Schematic of the experimental setup in optical tweezer experiments. (B) Schematic representation of hairpin DNA construction for single-molecule experiments. (C) Gel shift assay showing the success of making the hairpin construct. Lane 1, DNA ladder; lane 2, DNA handle mix; lane 3, ligated product, the ligated hairpin construct with handles on both sides is indicated with a red arrow.

Materials

- **Equipment**
  MiniTweezers (custom-made)
  PCR instrument (T100 Thermal Cycler, Bio-Rad Life Sciences Co., Ltd)
  NanoDrop 2000 Spectrophotometer (Thermo Fisher Scientific)
  Horizontal Electrophoresis System (Bio-Rad Life Sciences Co., Ltd)
- **Consumables**
  Polystyrene beads: 1% w/v antidigoxigenin antibody-coated polystyrene particles (3.0 μm, named DIG beads), streptavidin coated polystyrene particles (2.1 μm, named SA beads), both from Spherotech, Inc, IL.
  Customized glass tubing (King Precision Glass, Inc., CA): Bead dispenser tubes with 100 μm outer diameter (OD) and 25 μm inner diameter (ID).
  Customized cover glass (60 mm × 24 mm × 0.2 mm) (WaferPlus Technology Co. Ltd)
  Plastic tubing: Polyethylene tubing PE10, PE50 (Becton Dickinson).
  Nescofilm (Karlan research).
- **Reagents**
  Sodium chloride (NaCl, Sigma-Aldrich)
  4-(2-Hydroxyethyl)-1-piperazine-ethanesulfonic acid (HEPES) (Sigma-Aldrich)
  Pfu DNA polymerase DNA polymerase (Sangon)
  dNTP mix (10 mM, Invitrogen)
  PCR purification kit (Qiagen)
  Gel Purification Kit (Qiagen)
  T4 DNA ligase (Invitrogen)
  pBR322 vector (NEB)

Methods

- **Step 1: DNA Handle Preparation**
  Two types of dsDNA handles (2,000 bp) were prepared by means of Autosticky PCR [111] using pBR322 vector DNA as the template, one digoxigenin labeled, the other biotin-labeled (Figure 2.9B). The DNA handles contains a 15 nucleotide (nt) overhang at one end, which is introduced through a dspacer in the PCR primer.

In a PCR tube, prepare a 50 μL reaction mixture containing 5 μL PCR buffer (10 μL), 1 μL 10 mM dNTP mix, 0.5 μL 100 μM forward primer, 0.5 μL, 100 μM reverse primer, 20 ng DNA template, 0.5 μL Pfu DNA Polymerase. Add H_2O to a final volume of 50 μL. The PCR amplification is performed as described in the product manual. The PCR product should then be purified using a PCR purification kit.

- **Step 2: Preparation of a Hairpin DNA Construct**

  The hairpin DNA construct used in our optical tweezers experiments was prepared through the ligation of one oligonucleotide (contains a 30 bp hairpin) to the two DNA handles as in Figure 2.9B. For each ligation reaction, 2 μg of DNA handles is used, and the amount of an oligonucleotide linker is derived based on a linker to handle molar ratio of 5:1 in the final ligation reaction. The ligated sample could be tested directly using 1% agarose gel electrophoresis (Figure 2.9C). The final product was further gel-purified using the gel purification kit, and kept in the freezer before use.

- **Step 3: Microfluidic Chamber Preparation**

  We use a home-made microfluidic chamber in our optical tweezers experiments. The chamber is formed by sandwiching two coverslips with Nescofilms cut into specific shapes to form three channels: the top and bottom channels for bead injection and the central channel for optical trapping (Figure 2.10A). The top and bottom channels are connected to the central channel through dispenser glass tubing. The chamber is affixed to a customized chamber frame, connected to plastic tubing (Figure 2.10B), and installed on a motorized stage.

A B

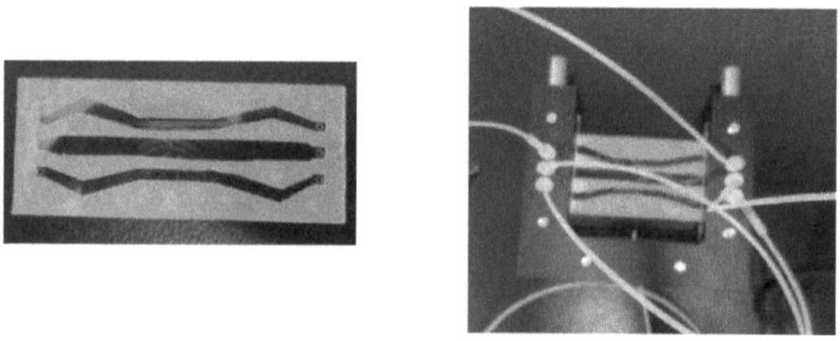

Figure 2.10 Microfluidic chamber used in optical tweezers experiments. (A) Microfluidic chamber showing three channels, two dispenser tubes, and the micropipette tube. The thickness of channels between two coverslips is ~200 μm. (B) Microfluidic chamber installed on a motorized translational stage and positioned between two objectives. The PE tubing connected to the three channels can be seen.

- **Step 4: Single-Molecule DNA Unzipping Using Optical Tweezers**

  4.1 Clean the microfluidic chamber and the tubes thoroughly by washing all the three channels with >10 mL HEPES buffer (20 mM HEPES, pH 7.5, 150 mM NaCl).

  4.2 Binding the DNA hairpin construct to the Dig beads. Mix around 100 ng DNA sample with 10 μL DIG beads and incubate on ice for 15 min. Dilute the beads into 2 mL HEPES buffer and transfer the beads to a 5 mL syringe.

  4.3 Dilute 10 μL of SA beads into 2 mL HEPES buffer and transfer the beads to a 5 mL syringe.

  4.4 Connect the syringes to the three channels in the microfluidic chamber through the washed PE tubing and minimize the flow in all three channels.

  4.5 Catch beads. Slowly inject the SA beads to the top or bottom channel until the beads diffuse from the tips of the dispenser tube into the central channel. Move the chamber stage to position the trap near the tip of glass tubing connected to the top channel and catch a single SA bead, and place it on the tip end of the micropipette. The SA beads could be sucked onto the tip end. Similarly catch a single DIG bead from the other channel and move it close to the SA bead.

  4.6 Form a hairpin DNA tether. Move the DIG beads close to each other to allow the biotinylated DNA handle to bind streptavidin on the SA bead surface and then slowly separate the two beads to check if the force between the two beads increases upon their separation. If the force increases, a tether is formed between the two beads. Otherwise, repeat the above approach-separation cycle until a tether is formed. Alternatively, test a different pair of beads.

  4.7 Once a tether is formed, the DNA hairpin could be stretched and relaxed multiple times by moving the pipette relative to the optical trap. In our instrumental setup, the applied force is determined by measuring the change in light momentum of the beams leaving the optical trap, while the extension of the molecule is determined by means of a "light lever system." Under tension, the unfolding of a DNA hairpin is accompanied by a large change in its extension, as the molecule goes from a compact native state to an elongated unfolded state. On the contrary, the refolding process causes a sharp compaction of the hairpin. These sudden changes in molecular end-to-end distance give rise to sharp transitions in the force vs. extension curves (Figure 2.11A).

4.8 At the end of the experiment, wash all three channels with 20% alcohol and then seal the channels in this solution to prevent bacterial growth in the microfluidic chamber.

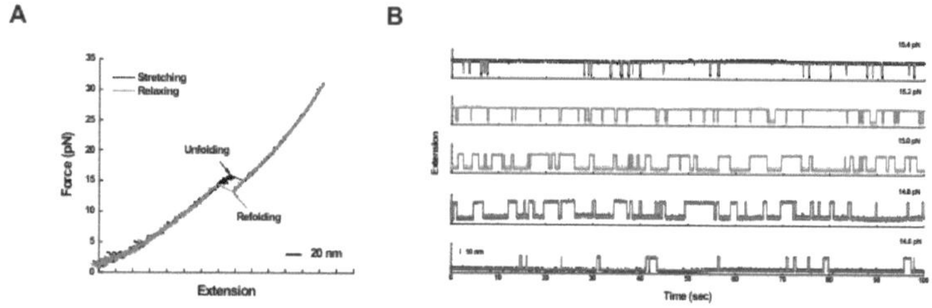

Figure 2.11 Mechanical folding and unfolding of single DNA hairpin by optical tweezers. (A) FEC obtained by stretching and relaxing a 30 bp DNA hairpin molecule. The hairpin unfolds at about 15 pN. (B) Extension vs. time trace showing an RNase H*Q4C/V155C protein hopping between its unfolded and folded states when holding the protein at a constant force of 14.6 pN–15.4 pN. The extension values are the distances between the centers of the tethering beads. The red lines are the idealized extension trajectories derived from hidden Markov modeling.

- **Step 5: Data Analysis**

To measure the energies of the states involved in the hairpin DNA folding processes and their associated transition kinetics as a function of force and to identify the possible intermediates in hairpin folding, FECs and the time-dependent trajectories must be analyzed in detail. The FECs provide a complete phase diagram of hairpin folding and unfolding transitions over a large force range (Figure 2.11A). The time-dependent trajectories at different constant forces allow us to determine protein folding energy and kinetics under equilibrium conditions in much greater detail. Figure 2.11B shows the time-extension trajectories of the transitions of the 30 bp DNA hairpin in the force region of 14 pN–16 pN. These single-molecule trajectories contain information on both the thermodynamics and kinetics of protein folding, which are the major subject of the data analysis. The hidden Markov modeling could be used to analyze the trajectories and determine hairpin folding energy and kinetics [75, 112].

References

1. Bustamante C, Chemla YR, Forde NR, Izhaky D (2004). Mechanical processes in biochemistry, *Annu Rev Biochem*, **73**, 705–748.
2. Ingber DE (1997). Tensegrity: the architectural basis of cellular mechanotransduction, *Annu Rev Physiol*, **59**, 575–599.
3. Ashkin A, Dziedzic JM, Bjorkholm JE, Chu S (1986). Observation of a single-beam gradient force optical trap for dielectric particles, *Opt Lett*, **11**, 288–290.
4. Asplund MC, Johnson JA, Patterson JE (2019). The 2018 nobel prize in physics: optical tweezers and chirped pulse amplification, *Anal Bioanal Chem*, **411**, 5001–5005.
5. Nieminen TA, du Preez-Wilkinson N, Stilgoe AB, Loke VLY, Bui AAM, Rubinsztein-Dunlop H (2014). Optical tweezers: theory and modelling, *J Quant Spectrosc Ra*, **146**, 59–80.
6. Rawson EG, May AD (1966). Propulsion and angular stabilization of dust particles in a laser cavity. *Appl Phys Lett*, **8**, 93–95.
7. May AD, Rawson EG, Hara EH (1967). Propulsion and angular stabilization of dust particles in a laser cavity II, *J Appl Phys*, **38**, 5290–5292.
8. Ashkin A (1970). Acceleration and trapping of particles by radiation pressure, *Phys Rev Lett*, **24**, 156–159.
9. Ashkin A, Schutze K, Dziedzic JM, Euteneuer U, Schliwa M (1990). Force generation of organelle transport measured in vivo by an infrared laser trap, *Nature*, **348**, 346–348.
10. Svoboda K, Block SM (1994). Biological applications of optical forces, *Annu Rev Biophys Biomol Struct*, **23**, 247–285.
11. Neuman KC, Block SM (2004). Optical trapping, *Rev Sci Instrum*, **75**, 2787–2809.
12. Quidant R, Petrov D, Badenes G (2005). Radiation forces on a Rayleigh dielectric sphere in a patterned optical near field, *Opt Lett*, **30**, 1009–1011.
13. Neto PAM, Nussenzveig HM (2000). Theory of optical tweezers, *Europhys Lett*, **50**, 702–708.
14. Rohrbach A (2005). Stiffness of optical traps: quantitative agreement between experiment and electromagnetic theory, *Phys Rev Lett*, **95**.

15. Ashkin A (1998). Forces of a single-beam gradient laser trap on a dielectric sphere in the ray optics regime, *Method Cell Biol*, **55**, 1–27.

16. Mazolli A, Neto PAM, Nussenzveig HM (2003). Theory of trapping forces in optical tweezers, *P Roy Soc a-Math Phy*, **459**, 3021–3041.

17. Dutra RS, Viana NB, Neto PAM, Nussenzveig HM (2014). Absolute calibration of forces in optical tweezers, *Phys Rev A*, **90**.

18. Dutra RS, Viana NB, Neto PAM, Nussenzveig HM (2016). Absolute calibration of optical tweezers: the MDSA plus theory, *Proc Spie*, **9922**.

19. Smith SB, Cui YJ, Bustamante C (2003). Optical-trap force transducer that operates by direct measurement of light momentum, *Method Enzymol*, **361**, 134–162.

20. Moffitt JR, Chemla YR, Izhaky D, Bustamante C (2006). Differential detection of dual traps improves the spatial resolution of optical tweezers, *Proc Natl Acad Sci U S A*, **103**, 9006–9011.

21. Bustamante C, Chemla YR, Moffitt JR (2009). High-resolution dual-trap optical tweezers with differential detection: an introduction, *Cold Spring Harb Protoc*, **2009**, pdb top60.

22. Bustamante C, Chemla YR, Moffitt JR (2009). High-resolution dual-trap optical tweezers with differential detection: alignment of instrument components, *Cold Spring Harb Protoc*, **2009**, pdb ip76.

23. Bustamante C, Chemla YR, Moffitt JR (2009). High-resolution dual-trap optical tweezers with differential detection: minimizing the influence of measurement noise, *Cold Spring Harb Protoc*, **2009**, pdb ip75.

24. Bustamante C, Chemla YR, Moffitt JR (2009). High-resolution dual-trap optical tweezers with differential detection: data collection and instrument calibration, *Cold Spring Harb Protoc*, **2009**, pdb ip74.

25. Bustamante C, Chemla YR, Moffitt JR (2009). High-resolution dual-trap optical tweezers with differential detection: instrument design, *Cold Spring Harb Protoc*, **2009**, pdb ip73.

26. Bustamante C, Chemla YR, Moffitt JR (2009). High-resolution dual-trap optical tweezers with differential detection: managing environmental noise, *Cold Spring Harb Protoc*, **2009**, pdb ip72.

27. Neuman KC, Chadd EH, Liou GF, Bergman K, Block SM (1999). Characterization of photodamage to *Escherichia coli* in optical traps, *Biophys J*, **77**, 2856–2863.

28. Tolic-Norrelykke SF, Schaffer E, Howard J, Pavone FS, Julicher F, Flyvbjerg H (2006). Calibration of optical tweezers with positional detection in the back focal plane, *Rev Sci Instrum*, **77**.

29. Wuite GJ, Davenport RJ, Rappaport A, Bustamante C (2000). An integrated laser trap/flow control video microscope for the study of single biomolecules, *Biophys J*, **79**, 1155–1167.

30. Keller M, Schilling J, Sackmann E (2001). Oscillatory magnetic bead rheometer for complex fluid microrheometry, *Rev Sci Instrum*, **72**, 3626–3634.

31. Gittes F, Schmidt CF (1998). Interference model for back-focal-plane displacement detection in optical tweezers, *Opt Lett*, **23**, 7–9.

32. Felgner H, Muller O, Schliwa M (1995). Calibration of light forces in optical tweezers, *Appl Opt*, **34**, 977–982.

33. Berg-Sorensen K, Flyvbjerg H (2004). Power spectrum analysis for optical tweezers, *Rev Sci Instrum*, **75**, 594–612.

34. Berg-Sorensen K, Peterman EJG, Weber T, Schmidt CF, Flyvbjerg H (2006). Power spectrum analysis for optical tweezers. II: laser wavelength dependence of parasitic filtering, and how to achieve high bandwidth, *Rev Sci Instrum*, **77**.

35. Ishijima A, Kojima H, Funatsu T, et al. (1998). Simultaneous observation of individual ATPase and mechanical events by a single myosin molecule during interaction with actin, *Cell*, **92**, 161–171.

36. Lang MJ, Fordyce PM, Engh AM, Neuman KC, Block SM (2004). Simultaneous, coincident optical trapping and single-molecule fluorescence, *Nat Methods*, **1**, 133–139.

37. Hohng S, Zhou R, Nahas MK, et al. (2007). Fluorescence-force spectroscopy maps two-dimensional reaction landscape of the holliday junction, *Science*, **318**, 279–283.

38. Comstock MJ, Ha T, Chemla YR (2011). Ultrahigh-resolution optical trap with single-fluorophore sensitivity, *Nat Methods*, **8**, 335–340.

39. Zhou RB, Kozlov AG, Roy R, et al. (2011). SSB functions as a sliding platform that migrates on DNA via reptation, *Cell*, **146**, 222–232.

40. Sirinakis G, Ren Y, Gao Y, Xi Z, Zhang Y (2012). Combined versatile high-resolution optical tweezers and single-molecule fluorescence microscopy, *Rev Sci Instrum*, **83**, 093708.

41. Comstock MJ, Whitley KD, Jia H, et al. (2015). Protein structure. Direct observation of structure-function relationship in a nucleic acid-processing enzyme, *Science*, **348**, 352–354.

42. Ngo TTM, Zhang QC, Zhou RB, Yodh JG, Ha T (2015). Asymmetric unwrapping of nucleosomes under tension directed by DNA local flexibility, *Cell*, **160**, 1135–1144.

43. Reihani SN, Oddershede LB (2007). Optimizing immersion media refractive index improves optical trapping by compensating spherical aberrations, *Opt Lett*, **32**, 1998–2000.
44. Dholakia K, Cizmar T (2011). Shaping the future of manipulation, *Nat Photonics*, **5**, 335–342.
45. Novotny L, Bian RX, Xie XS (1997). Theory of nanometric optical tweezers, *Phys Rev Lett*, **79**, 645–648.
46. Grigorenko AN, Roberts NW, Dickinson MR, Zhang Y (2008). Nanometric optical tweezers based on nanostructured substrates, *Nat Photonics*, **2**, 365–370.
47. Righini M, Ghenuche P, Cherukulappurath S, Myroshnychenko V, de Abajo FJG, Quidant R (2009). Nano-optical trapping of Rayleigh particles and *Escherichia coli* bacteria with resonant optical antennas, *Nano Lett*, **9**, 3387–3391.
48. Zhang WH, Huang LN, Santschi C, Martin OJF (2010). Trapping and sensing 10 nm metal nanoparticles using plasmonic dipole antennas, *Nano Lett*, **10**, 1006–1011.
49. Juan ML, Righini M, Quidant R (2011). Plasmon nano-optical tweezers, *Nat Photonics*, **5**, 349–356.
50. Pang YJ, Gordon R (2011). Optical trapping of 12 nm dielectric spheres using double-nanoholes in a gold film, *Nano Lett*, **11**, 3763–3767.
51. Berthelot J, Acimovic SS, Juan ML, Kreuzer MP, Renger J, Quidant R (2014). Three-dimensional manipulation with scanning near-field optical nanotweezers, *Nat Nanotechnol*, **9**, 295–299.
52. Bustamante C, Bryant Z, Smith SB (2003). Ten years of tension: single-molecule DNA mechanics, *Nature*, **421**, 423–427.
53. Greenleaf WJ, Woodside MT, Block SM (2007). High-resolution, single-molecule measurements of biomolecular motion, *Annu Rev Biophys Biomol Struct*, **36**, 171–190.
54. Larson MH, Greenleaf WJ, Landick R, Block SM (2008). Applied force reveals mechanistic and energetic details of transcription termination, *Cell*, **132**, 971–982.
55. Bustamante C, Cheng W, Meija YX (2011). Revisiting the central dogma one molecule at a time, *Cell*, **144**, 480–497.
56. Bustamante CJ, Kaiser CM, Maillard RA, Goldman DH, Wilson CA (2014). Mechanisms of cellular proteostasis: insights from single-molecule approaches, *Annu Rev Biophys*, **43**, 119–140.
57. Smith SB, Cui Y, Bustamante C (1996). Overstretching B-DNA: the elastic response of individual double-stranded and single-stranded DNA molecules, *Science*, **271**, 795–799.

58. Baumann CG, Smith SB, Bloomfield VA, Bustamante C (1997). Ionic effects on the elasticity of single DNA molecules, *Proc Natl Acad Sci U S A*, **94**, 6185–6190.

59. Wang MD, Yin H, Landick R, Gelles J, Block SM (1997). Stretching DNA with optical tweezers, *Biophys J*, **72**, 1335–1346.

60. Bosaeus N, El-Sagheer AH, Brown T, et al. (2012). Tension induces a base-paired overstretched DNA conformation, *Proc Natl Acad Sci U S A*, **109**, 15179–15184.

61. Woodside MT, Block SM (2014). Reconstructing folding energy landscapes by single-molecule force spectroscopy, *Annu Rev Biophys*, **43**, 19–39.

62. Dudko OK, Hummer G, Szabo A (2006). Intrinsic rates and activation free energies from single-molecule pulling experiments, *Phys Rev Lett*, **96**, 108101.

63. Dudko OK (2009). Single-molecule mechanics: new insights from the escape-over-a-barrier problem, *Proc Natl Acad Sci U S A*, **106**, 8795–8796.

64. Suzuki Y, Dudko OK (2010). Single-molecule rupture dynamics on multidimensional landscapes, *Phys Rev Lett*, **104**, 048101.

65. Gupta AN, Vincent A, Neupane K, Yu H, Wang F, Woodside MT (2011). Experimental validation of free-energy-landscape reconstruction from non-equilibrium single-molecule force spectroscopy measurements, *Nat Phys*, **7**, 631–634.

66. Pierse CA, Dudko OK (2013). Kinetics and energetics of biomolecular folding and binding, *Biophys J*, **105**, L19–L22.

67. Jarzynski C (1997). Nonequilibrium equality for free energy differences, *Phys Rev Lett*, **78**, 2690–2693.

68. Liphardt J, Dumont S, Smith SB, Tinoco I, Bustamante C (2002). Equilibrium information from nonequilibrium measurements in an experimental test of Jarzynski's equality, *Science*, **296**, 1832–1835.

69. Bell GI (1978). Models for specific adhesion of cells to cells, *Science*, **200**, 618–627.

70. Tinoco I, Jr, Li PT, Bustamante C (2006). Determination of thermodynamics and kinetics of RNA reactions by force, *Q Rev Biophys*, **39**, 325–360.

71. Liphardt J, Onoa B, Smith SB, Tinoco I, Jr, Bustamante C (2001). Reversible unfolding of single RNA molecules by mechanical force, *Science*, **292**, 733–737.

72. Cecconi C, Shank EA, Bustamante C, Marqusee S (2005). Direct observation of the three-state folding of a single protein molecule, *Science*, **309**, 2057–2060.

73. Woodside MT, Anthony PC, Behnke-Parks WM, Larizadeh K, Herschlag D, Block SM (2006). Direct measurement of the full, sequence-dependent folding landscape of a nucleic acid, *Science*, **314**, 1001–1004.

74. Greenleaf WJ, Frieda KL, Foster DAN, Woodside MT, Block SM (2008). Direct observation of hierarchical folding in single riboswitch aptamers, *Science*, **319**, 630–633.

75. Stigler J, Ziegler F, Gieseke A, Gebhardt JC, Rief M (2011). The complex folding network of single calmodulin molecules, *Science*, **334**, 512–516.

76. Schliwa M, Woehlke G (2003). Molecular motors, *Nature*, **422**, 759–765.

77. Kolomeisky AB (2013). Motor proteins and molecular motors: how to operate machines at the nanoscale, *J Phys-Condens Mat*, **25**.

78. Block SM (2007). Kinesin motor mechanics: binding, stepping, tracking, gating, and limping, *Biophys J*, **92**, 2986–2995.

79. Block SM, Goldstein LSB, Schnapp BJ (1990). Bead movement by single kinesin molecules studied with optical tweezers, *Nature*, **348**, 348–352.

80. Svoboda K, Schmidt CF, Schnapp BJ, Block SM (1993). Direct observation of kinesin stepping by optical trapping interferometry, *Nature*, **365**, 721–727.

81. Asbury CL, Fehr AN, Block SM (2003). Kinesin moves by an asymmetric hand-over-hand mechanism, *Science*, **302**, 2130–2134.

82. Altman D, Sweeney HL, Spudich JA (2004). The mechanism of myosin VI translocation and its load-induced anchoring, *Cell*, **116**, 737–749.

83. Okten Z, Churchman LS, Rock RS, Spudich JA (2004). Myosin VI walks hand-over-hand along actin, *Nat Struct Mol Biol*, **11**, 884–887.

84. Rai AK, Rai A, Ramaiya AJ, Jha R, Mallik R (2013). Molecular adaptations allow dynein to generate large collective forces inside cells, *Cell*, **152**, 172–182.

85. Schnitzer MJ, Visscher K, Block SM (2000). Force production by single kinesin motors, *Nat Cell Biol*, **2**, 718–723.

86. Smith DE, Tans SJ, Smith SB, Grimes S, Anderson DL, Bustamante C (2001). The bacteriophage straight phi29 portal motor can package DNA against a large internal force, *Nature*, **413**, 748–752.

87. Chemla YR, Aathavan K, Michaelis J, et al. (2005). Mechanism of force generation of a viral DNA packaging motor, *Cell*, **122**, 683–692.

88. Aathavan K, Politzer AT, Kaplan A, et al. (2009). Substrate interactions and promiscuity in a viral DNA packaging motor, *Nature*, **461**, 669–673.

89. Chistol G, Liu S, Hetherington CL, et al. (2012). High degree of coordination and division of labor among subunits in a homomeric ring ATPase, *Cell*, **151**, 1017–1028.

90. Liu S, Chistol G, Hetherington CL, et al. (2014). A viral packaging motor varies its DNA rotation and step size to preserve subunit coordination as the capsid fills, *Cell*, **157**, 702–713.

91. Yin H, Wang MD, Svoboda K, Landick R, Block SM, Gelles J (1995). Transcription against an applied force, *Science*, **270**, 1653–1657.

92. Wang MD, Schnitzer MJ, Yin H, Landick R, Gelles J, Block SM (1998). Force and velocity measured for single molecules of RNA polymerase, *Science*, **282**, 902–907.

93. Davenport RJ, Wuite GJ, Landick R, Bustamante C (2000). Single-molecule study of transcriptional pausing and arrest by *E. coli* RNA polymerase, *Science*, **287**, 2497–2500.

94. Neuman KC, Abbondanzieri EA, Landick R, Gelles J, Block SM (2003). Ubiquitous transcriptional pausing is independent of RNA polymerase backtracking, *Cell*, **115**, 437–447.

95. Shaevitz JW, Abbondanzieri EA, Landick R, Block SM (2003). Backtracking by single RNA polymerase molecules observed at near-base-pair resolution, *Nature*, **426**, 684–687.

96. Abbondanzieri EA, Greenleaf WJ, Shaevitz JW, Landick R, Block SM (2005). Direct observation of base-pair stepping by RNA polymerase, *Nature*, **438**, 460–465.

97. Herbert KM, La Porta A, Wong BJ, et al. (2006). Sequence-resolved detecton of pausing by single RNA polymerase molecules, *Cell*, **125**, 1083–1094.

98. Hodges C, Bintu L, Lubkowska L, Kashlev M, Bustamante C (2009). Nucleosomal fluctuations govern the transcription dynamics of RNA polymerase II, *Science*, **325**, 626–628.

99. Larson MH, Mooney RA, Peters JM, et al. (2014). A pause sequence enriched at translation start sites drives transcription dynamics in vivo, *Science*, **344**, 1042–1047.

100. Fazal FM, Meng CA, Murakami K, Kornberg RD, Block SM (2015). Real-time observation of the initiation of RNA polymerase II transcription, *Nature*, **525**, 274-+.

101. Righini M, Lee A, Canari-Chumpitaz C, et al. (2018). Full molecular trajectories of RNA polymerase at single base-pair resolution, *P Natl Acad Sci USA*, **115**, 1286–1291.

102. Wen JD, Lancaster L, Hodges C, et al. (2008). Following translation by single ribosomes one codon at a time, *Nature*, **452**, 598–603.

103. Kaiser CM, Goldman DH, Chodera JD, Tinoco I, Jr, Bustamante C (2011). The ribosome modulates nascent protein folding, *Science*, **334**, 1723–1727.

104. Qu X, Wen JD, Lancaster L, Noller HF, Bustamante C, Tinoco I, Jr (2011). The ribosome uses two active mechanisms to unwind messenger RNA during translation, *Nature*, **475**, 118–121.

105. Qu X, Lancaster L, Noller HF, Bustamante C, Tinoco I, Jr (2012). Ribosomal protein S1 unwinds double-stranded RNA in multiple steps, *Proc Natl Acad Sci U S A*, **109**, 14458–14463.

106. Kim HK, Liu F, Fei J, et al. (2014). A frameshifting stimulatory stem loop destabilizes the hybrid state and impedes ribosomal translocation, *Proc Natl Acad Sci U S A*, **111**, 5538–5543.

107. Goldman DH, Kaiser CM, Milin A, Righini M, Tinoco I, Jr, Bustamante C (2015). Ribosome. Mechanical force releases nascent chain-mediated ribosome arrest in vitro and in vivo, *Science*, **348**, 457–460.

108. Yan S, Wen JD, Bustamante C, Tinoco I, Jr (2015). Ribosome excursions during mRNA translocation mediate broad branching of frameshift pathways, *Cell*, **160**, 870–881.

109. Alexander LM, Goldman DH, Wee LM, Bustamante C (2019). Non-equilibrium dynamics of a nascent polypeptide during translation suppress its misfolding, *Nat Commun*, **10**, 2709.

110. Desai VP, Frank F, Lee A, et al. (2019). Co-temporal force and fluorescence measurements reveal a ribosomal gear shift mechanism of translation regulation by structured mRNAs, *Mol Cell*, **75**, 1007–1019 e1005.

111. Gal J, Schnell R, Szekeres S, Kalman M (1999). Directional cloning of native PCR products with preformed sticky ends (Autosticky PCR), *Mol Gen Genet*, **260**, 569–573.

112. Zhang YL, Jiao JY, Rebane AA (2016). Hidden Markov modeling with detailed balance and its application to single protein folding, *Biophys J*, **111**, 2110–2124.

Chapter 3

Single-Molecule Biosensing by Fluorescence Resonance Energy Transfer

Ying Lu, Jianbing Ma, and Ming Li
Beijing National Laboratory for Condensed Matter Physics,
Institute of Physics, Chinese Academy of Sciences, Beijing 100190, China
yinglu@iphy.ac.cn

The past three decades have seen the rapid expansion of the fluorescence resonance energy transfer (FRET) method in the single-molecule region. With its recent progress in high precision and high-throughput measurements, single-molecule FRET (smFRET) has become one of the most prominent tools with which to trace real-time dynamics and conformational changes of biomolecules in their active states. In this chapter, we first provide a brief introduction of the basic concept of smFRET and then we review its important applications. Subsequently, we summarize recent efforts in overcoming the shortcomings of conventional smFRET methods. In an Appendix, we provide a tutorial protocol, which aims to provide the readers a hands-on experience to measure single-molecule dynamics of complex DNA structures.

Single-Molecule Tools for Bioanalysis
Edited by Shuo Huang

ISBN 978-981-4800-44-0 (Hardcover), 978-1-003-18913-8 (eBook)
www.jennystanford.com

3.1 Introduction

FRET is a quantum mechanical process in which direct nonradiation transfer of energy from a donor to an acceptor occurs by means of intermolecular long-range dipole–dipole coupling (Figure 3.1). FRET was first identified in the 1920s by Cario, Franck, and Perrin [1–4]. In the 1940s, a quantitative FRET theory was formulated independently by Förster and Oppenheimer [5]. Subsequently, the concept of FRET was widely adapted to measurements of subtle distance changes between donor and acceptor fluorophores, which are labeled on target sites of an analyte to investigate molecular dynamics and reactions in ensembles, both in vitro and in vivo [6]. Though it excels with a measurement dynamic range of 1–10 nm [7, 8], results acquired by FRET at this moment normally yield averages over a large number of molecules and events.

Biologists however desire to elucidate every molecular events that occur in living systems [9]. Unfortunately, conformational transitions and dynamics of biological molecules are not synchronized with each other and are therefore difficult to be detected by FRET in bulk, if not resolvable in single molecule. The development of near-field optical microscopy in the 1990s enabled the detection of an individual fluorophore at room temperature [10–13], which has assisted the first demonstration of smFRET measurement [14–17]. Since then, smFRET has become one of the leading techniques for studies of interactions and dynamics of biomolecules [17–23]. The process has also been widely applied to investigation of dynamics in heterogeneous systems and transient conformational changes in nano-systems [5, 24, 25]. It assisted in the unraveling of the molecular mechanisms underlying the dynamics of nucleic acids as well as interactions between nucleic acids and proteins, including DNA replications, repair, transcriptions and translations, to name a few [5, 26–33].

Experimentally, for FRET to occur, the following conditions must be simultaneously fulfilled [34, 35]. (i) The emission spectrum of the donor must significantly overlap with the absorption spectrum of the acceptor. (ii) The distance between the donor and the acceptor must fall within the operative range of FRET. (iii) The donor emission dipole moment, the acceptor absorption dipole moment and their separation vectors must be in a favorable mutual orientation. (iv) The donor should have a high quantum yield [36].

The transfer efficiency of smFRET follows the equation of $E = 1/(1 + [R/R_0]^6)$, in which R is the distance between the donor and the acceptor. R_0 is the distance at which 50% of the energy is transferred (Figure 3.1c) and is determined by the opto-physical properties of the fluorophores and the relative orientation of their dipole moments.

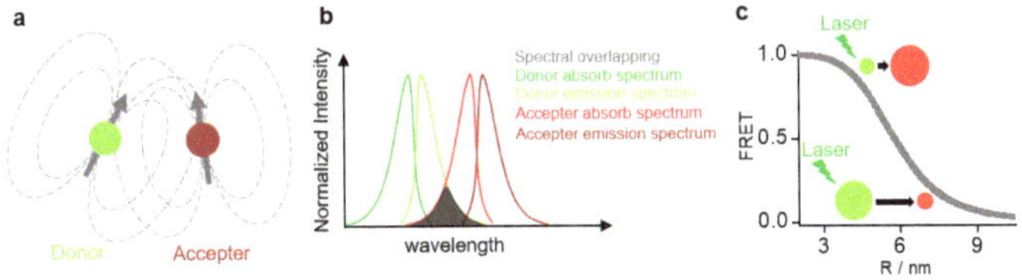

Figure 3.1 Principle of FRET. **(a)** Scheme of dipole–dipole interaction between a donor and an acceptor. **(b)** Overlap of the donor emission spectrum and the acceptor absorption spectrum. **(c)** The FRET efficiency as a function of distance between the donor and the acceptor.

Many advanced forms of FRET techniques have been recently developed. Multicolor FRET for example was developed to assess three-dimensional changes in distances between labeling dyes [37, 38]. Nanotensioner-enhanced FRET was invented to enhance the precision of DNA-related processes [19]. Photoactivation FRET (PAFRET) was proposed to break the concentration barrier of dyes in FRET [39] and the surface-induced fluorescence attenuation (SIFA) and the quenchers-in-a-liposome FRET (lipoFRET) were developed for the detection of macromolecules in lipid membranes [40, 41]. Detailed descriptions of these methods are available in relevant reviews [5, 6, 25, 36].

3.2 Implementation of smFRET

In order to conduct successful smFRET experiments, many aspects such as optical setup, labeling of fluorophores, measurement chamber fabrication and photo-protection strategy should be taken into consideration.

3.2.1 Optical Setup

A smFRET setup typically includes an inverted fluorescence microscope equipped with illumination lasers, fluorescence separation optics, and ultrasensitive photon detectors. Fluorescence from a single molecule is weak, often blurred by background noises originating from Raman and Rayleigh scatterings and from fluorescence of impurities in the solvent, coverslips, or optical components. Careful elimination of background fluorescence by various means, such as prebleaching of impurities in the solvent and the use of very low-fluorescing optical materials, can be applied [42]. Single-molecule detection also requires high-efficiency collection optics and sensitive detectors with a high quantum efficiency and low dark noises.

Specifically, according to the arrangement of illumination lasers and emission detectors, two types of FRET instrumentation, i.e., the total internal reflection (TIR) type (Figure 3.2a) and the confocal type (Figure 3.2b) [25, 43] are often adopted. The TIR type is useful for studies of immobilized molecules, allowing measurements of a single molecule in real time, while the confocal type, which has a higher temporal resolution, is good for measurement of moving molecules in liquid phases.

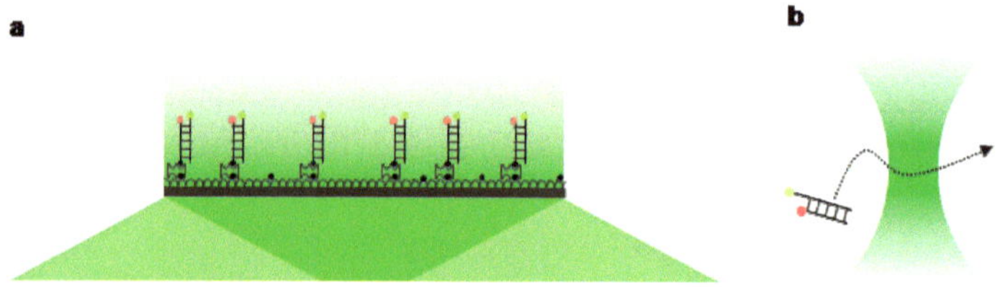

Figure 3.2 Two types of smFRET setup. **(a)** smFRET in a wide-field TIR light field. Hundreds of labeled molecules can be stimulated simultaneously. **(b)** smFRET in a confocal light field. Only the molecules in the small focal spot can be detected.

3.2.1.1 TIR-based setup

The TIR-based setup needs a TIR microscope in which an evanescent field of laser excitation is created on the upper surface of the glass coverslip [44]. This evanescent field extends only ~200 nm from the glass–liquid interface, and greatly reduces background fluorescence emitted from the liquid environment. Either of two procedures are usually adopted to generate the evanescent field (Figure 3.3) [45]. The first uses a high-numerical aperture (NA) oil immersion objective, with which TIR at the glass–liquid interface is achieved by adjusting the focus position of the laser beam to the back focal plane of the objective. This method has a high photon-collection efficiency and frees up the space above the sample for additional sample control [24]. The second, alternative method is to use a prism on top of the reaction chamber (Figure 3.3b). To reduce the background fluorescence, the fused silica prism and the slide are usually of choice. Oil with a similar refractive index between the silica prism and the slide was also used to lower the reflection at the interface of the slide and the prism. In this configuration, the laser beam is refracted by the prism, generating a TIR at the inner chamber surface. With either configuration, the fluorescence emission is collected by an objective beneath the coverslip and split by a dichroic mirror. Subsequent additional bandpass filters further reduce cross-talk between detection channels before the emission reaches the detectors [24]. Planar detectors, such as charge coupled devices or complementary metal oxide semiconductors, are commonly used and allow parallel detection of many immobile molecules.

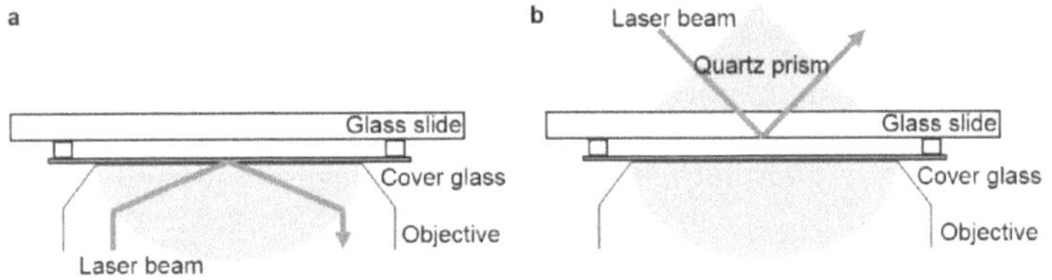

Figure 3.3 TIR-type smFRET. The evanescent field is created either by an NA objective (**a**) or a quartz prism (**b**).

3.2.1.2 Confocal-based setup

In a confocal-based setup, laser beams are focused in a volume of ~1 fL (10^{-15} L) to excite the fluorescent molecules. Either continuous lasers or pulsed lasers can be used as light sources. Point detectors, such as avalanche photodiodes or photomultiplier tubes, which have advantages of a high signal to background ratio, a microsecond temporal resolution and the ability to record complicated spectroscopies of mobile molecules, are often used in this configuration. The emitted fluorescence passes through a pinhole and is collected by a detector. With this configuration, the fluorescence out of the focused volume can hardly pass through the pinhole [46–49], reducing the interference from background noise. To extend the observation time, target biomolecules such as a Holliday junction, can be tethered to freely diffusing liposomes (Figure 3.4) [50].

Provided that the microscope stage can scan in the *x*–*y* plane, molecules can also be immobilized on the glass surface, and it is possible to identify the locations of individual molecules for direct visualization [51].

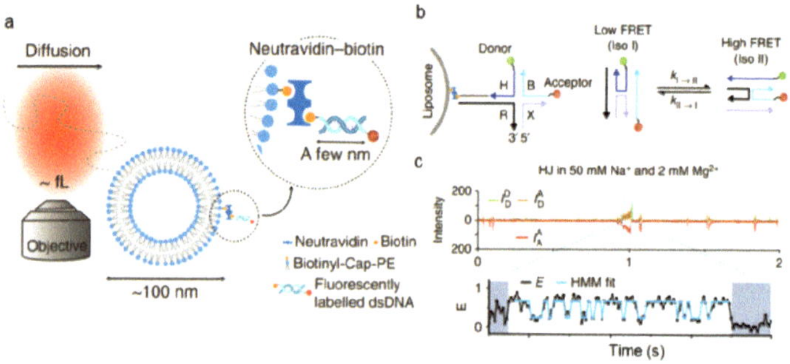

Figure 3.4 Confocal type smFRET. Reprinted with permission from ref. [50]. **(a)** Schematic description of the liposome-tethering approach designed to have long observation duration of a FRET molecule. Tethering of the FRET molecule to a liposome is achieved through the biotin-neutravidin interaction. **(b)** Schematic illustrations of the liposome tethering of a Holliday junction and the conformational transition of the Holliday junction. **(c)** Typical time trace of the Holliday junction dynamics at 50 mM NaCl and 2 mM $MgCl_2$. The FRET efficiency E (black line) was calculated for each time bin (0.5 ms). The blue line represents the fit to the FRET efficiencies obtained using hidden Markov modeling.

3.2.2 Fluorophore Labeling

Fluorophores used in smFRET are usually small organic dyes which introduce minimum perturbations to the structure of the host molecules under investigation. Fluorophores for smFRET should be photo-stable, strong emitters with low fluctuation of the emitted fluorescent intensity [52]. An ideal donor–acceptor pair in smFRET should have (i) a large spectral separation to minimize leakage of the donor emission into the acceptor channel and reduce the degree of direct excitation of the acceptor, (ii) a proper characteristic distance (R_0) between the donor and the acceptor, and (iii) comparable emission quantum yields to guarantee clearly anticorrelated intensity changes of the donor and the acceptor [25]. Many high-quality dyes are commercially available including the Cyanine family, the Alexa family, and the Atto family fluorophores, [38], among which the most widely used pair of FRET dyes are Cy3/Cy5 [19, 53–56].

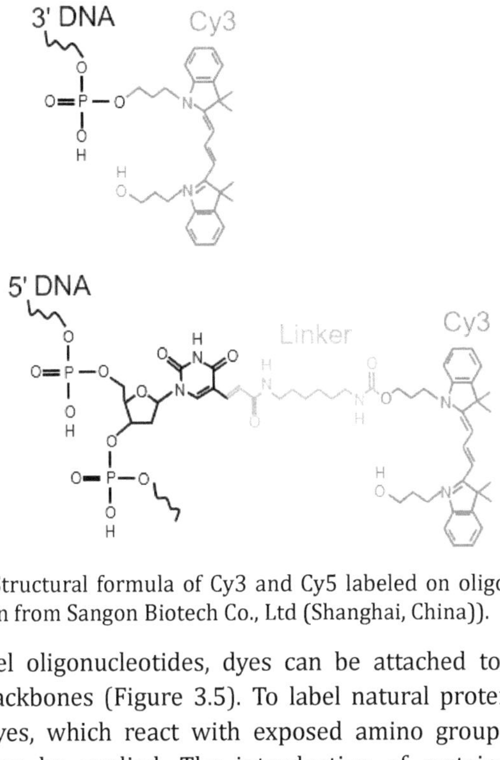

Figure 3.5 Structural formula of Cy3 and Cy5 labeled on oligonucleotides (Information from Sangon Biotech Co., Ltd (Shanghai, China)).

To label oligonucleotides, dyes can be attached to the bases or their backbones (Figure 3.5). To label natural proteins, amine-reactive dyes, which react with exposed amino groups of target proteins can be applied. The introduction of cysteine residues, which are not commonly present on the surfaces of natural proteins, allows for specific attachment of thiol-reactive fluorophores to the proteins [57, 58]. Many other methods have recently been developed for highly specific fluorophore labeling, including the introduction of noncanonical amino acids [59, 60], the attachment of a dye-conjugated tris-nitrilotriacetic acid (tris-NTA) to the histidine-tag of the protein [61] or the genetic insertion of small amino-acid sequences for specific attachment of small organic fluorophores

[62, 63]. Other efforts have also been used to label more than one fluorophore on the same biomolecule [17, 25, 64–66].

3.2.3 Surface Modification

To form an air-tight reaction chamber for smFRET measurements, a double-sided tape or a parafilm can be sandwiched between a precleaned glass slide and a coverslip (Figure 3.3). A pair of holes predrilled in the slide are used for exchanges of solutions by pipetting or pumping liquid in and out of the chamber [19, 24, 57]. Biomolecules under investigation may adhere nonspecifically to the inner surface of the reaction chamber. Thus, passivation of the surface is normally carried out by treatment of polyethylene glycol (PEG), bovine serum albumin (BSA), or Tween 20 [53, 67–70]. On the other side, biotin-labeled PEG or BSA serve to immobilize target biomolecules via the interaction of biotin and streptavidin. The PEG-coated system is a mixture of 99% mPEG (m-PEG-5000, Laysan Bio, Inc.) and 1% biotin-PEG (biotin-PEG-5000, Laysan Bio, Inc.). The biotinylated biomolecules are often added to the chamber with a ~100 pM final concentration to achieve a desired single-molecule density, which minimizes the chance of neighboring molecules overlapping [19, 24, 57, 71]. It is normally desirable that the observed spot density of background fluorescence should be <10% of the specifically tethered molecules. Alternatively, other antibody–antigen combinations can be applied in this manner for surface modification [72, 73].

3.2.4 Photo-Protection Strategy

Due to the formation of triplet states [74], fluorophores frequently undergo transient blinking during photon emissions and the use of triplet quenchers to reduce the blinking is recommended [19, 24, 74, 75]. In addition, high power laser illumination may result in photo-oxidation which bleaches the dyes, and a common strategy to reduce such photo-bleaching is to eliminate oxygen in the solution with an enzymatic oxygen scavenging system such as the glucose oxidation system (a mixture of glucose oxidase, catalase and β-D-glucose) [19, 24]. The use of a mixture of protocatechuic acid and protocatechuate-3,4-dioxygenase is an alternative option [24, 76].

3.3 Applications of smFRET

Currently, smFRET has become one of the most prevalent methods to study dynamic processes of biomolecules [25, 77, 78] and has found applications in many other situations [42, 54–56, 58, 79–86]. SmFRET can also detect rare conformation transitions or reaction intermediates that are difficult to be detected by other means. Here, we focus on applications of smFRET in studies of the dynamic structures of nucleic acids, proteins and nucleic acid-interacting enzymes.

3.3.1 Structural Dynamics of Nucleic Acids

Nucleic acids such as DNA and RNA are pivotal molecules during transcription, replication, and gene recombination. Many forms of DNA structures exist in cells, including B-DNA, Z-DNA, four-way Holliday junctions, and G-quadruplex DNA [87–93].

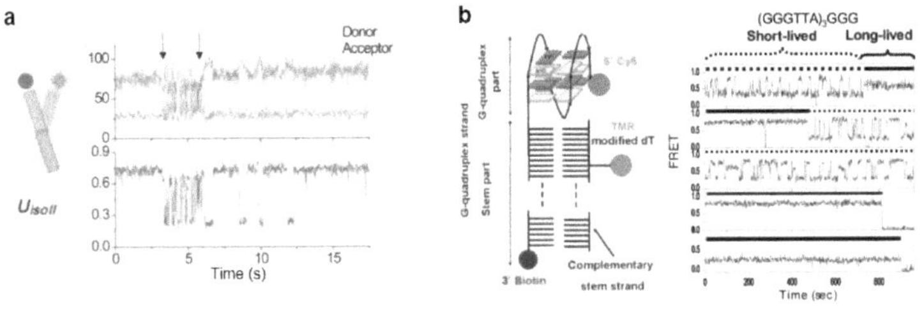

Figure 3.6 SmFRET study of DNA dynamics. **(a)** Spontaneous branch migration of two states in a Holliday junction. Reprinted with permission from ref. [54]. **(b)** Two different structure dynamic processes in a G-4 DNA. Reprinted with permission from ref. [55].

Genetic recombination between homologous DNA molecules results in transient formation of four-way Holliday junctions [94–96]. Once formed, the Holliday junction alone can undergo spontaneous branch migration. Using smFRET, McKinney et al. observed that the junction may switch between two distinct patterns, each consisting of two-state transitions but with markedly different kinetics [54] (Figure 3.6a).

G-quadruplexes are secondary structures formed by tandem repeats of guanines [97], and play important roles in regulating telomere lengths in vivo [98–102]. Using smFRET, Lee et al. probed the dynamics of G-quadruplexes in human telomeric DNA [55]. Three conformations, one unfolded and two folded, were observed in a solution containing potassium ions. Each conformation could be further divided into two species, the long-lived and the short-lived (Figure 3.6b). The long-lived states, which result from the folded structures, are dominant in physiologically relevant conditions. Although rare under these conditions, the short-lived species determine the overall dynamics because they bridge the different long-lived species. The extremely diverse conformations of the human telomeric DNA may have mechanistic implications for the proteins and drugs that recognize G-rich sequences [103].

Bending and loop formation of DNA shorter than 100 base pairs (bp) are common in cellular processes such as regulated gene expression in bacteria and eukaryotes [104, 105], packaging of DNA in viral capsids and DNA storage complexes in eukaryotes [106]. Quantifying the intrinsic bendability of DNA at these biologically important length scales is essential to an understanding of DNA–protein interactions. Vafabakhsh et al. developed a smFRET assay to study the cyclization of single DNA molecules in real time [56]. Between 67 and 106 bp, the looping rate has a weak length dependence that cannot be described by the widely used worm-like chain model [107]. Many biologically important protein–DNA interactions that involve looping and bending of DNA below 100 bp probably use this intrinsic bendability of DNA.

RNA plays a central role in cellular processes such as splicing and translation [108]. Ribozymes, which are RNA molecules that act as enzymes, have been recognized as ideal model systems with which to study the relationship of structure and function in RNA, because their catalytic activity directly reflects the extent of native structure formation [109–111]. Zhuang et al. studied the correlation between structural dynamics and function of the hairpin ribozyme [80]. They found complex structural dynamics with four docked states with distinct stabilities and a strong memory effect in which each molecule rarely switches between different docked states. They also found that the complex structural dynamics can quantitatively explain the heterogeneous cleavage kinetics common to many catalytic RNAs. The intimate coupling of structural dynamics and function is probably a general phenomenon for RNA. Subsequently, Rueda et al. quantified the effects of modifications of essential functional groups remote from the site of catalysis [81] (Figure 3.7). Their findings have broad implications for the action of drugs and ligands distal to the active site or the engineering of allostery into RNA [112, 113].

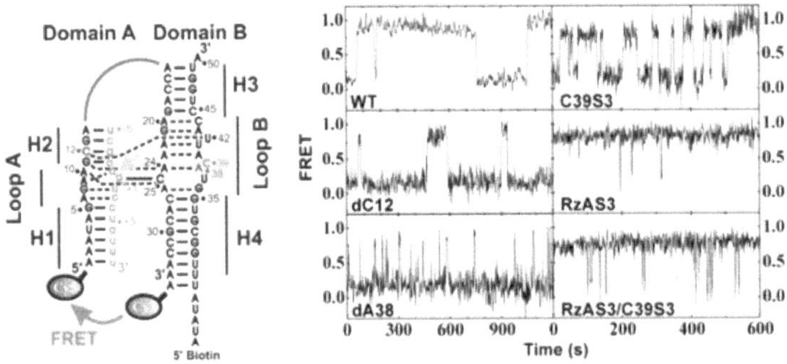

Figure 3.7 RNA enzymes dynamics. With several mutated hairpin ribozymes, it was found that essential functional groups impact catalysis from a distance. Reprinted with permission from ref. [81].

3.3.2 Protein Structural Dynamics

The functions of proteins are correlated with their 3D structures. Their native conformations are in many cases flexible and may be stabilized by a combination of kinetic and thermodynamic controls [114, 115]. SmFRET has provided new perspectives to study these fundamental processes [42, 58, 82]. For instance, the F_0F_1-ATP synthase is a membrane-bound enzyme that uses energy derived from an electrochemical proton gradient for ATP generation. With fluorophore-labeled F_0F_1-ATP synthases, Diez et al. observed directly the stepwise rotation of the subunit of the synthases during proton transport-powered ATP synthesis by smFRET, and observed three distinct distances to the b subunit in repeating sequences (Figure 3.8a) [58].

Many intrinsically disordered or unstructured proteins are associated with diseases, such as cancer or neurodegenerative disorders [116]. Such proteins often contain a large proportion of charged amino acids. Diez et al. investigated the influence of charged residues on the dimensions of unfolded and intrinsically disordered proteins [82]. Using smFRET, they found that three intrinsically disordered proteins, the N-terminal domain of HIV-1 integrase (IN), and human prothymosin α (ProTα) exhibit a prominent expansion at low ionic strength that can be correlated with their net charges (Figure 3.8b). The pronounced effect of charges on the dimensions of unfolded proteins has important implications for the cellular functions of intrinsically disordered proteins.

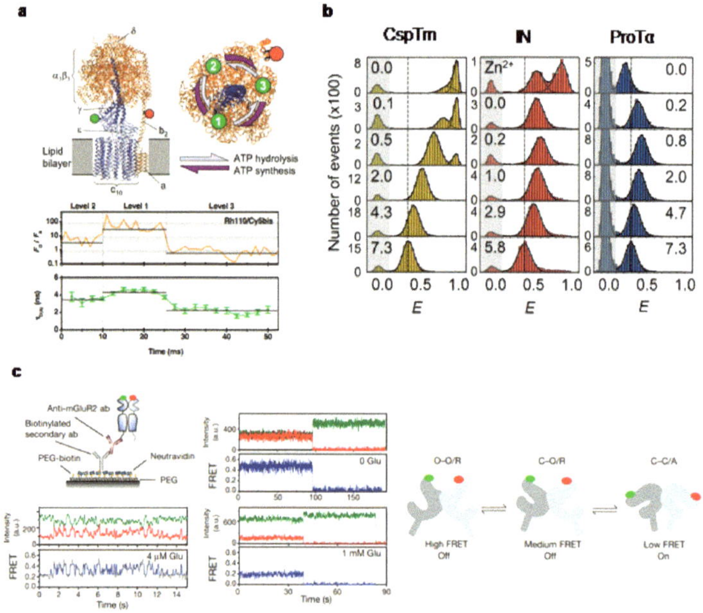

Figure 3.8 SmFRET studies of protein structural dynamics. **(a)** Rotation of the γ subunit proceeds in 120° steps (bottom panel) with donor on the "rotor" (γ subunit) and accepter on "stator" (b subunit). Reprinted with permission from ref. [58]. **(b)** GdmCl dependence of three proteins, the globular cold shock protein CspTm, the N-terminal domain of HIV-1 integrase (IN), and the human prothymosin α (ProTα). The molar GdmCl concentrations are indicated in each panel. Folded IN ($E \approx 0.9$) is only populated in the presence of $ZnCl_2$ (100 μM; 0 M GdmCl). Reprinted with permission from ref. [82]. **(c)** Three dynamic states of mGluR2 were observed at different Glu concentrations. A three-state fit obtained from hidden Markov model was showed with the filtered raw data at 4 μM Glu. Reprinted with permission from ref. [42].

SmFRET has also been applied to studies of the conformational dynamics of activation of G-protein-coupled receptors (GPCR) which constitute the largest family of membrane receptors in eukaryotes. Metabotropic glutamate receptors (mGluR) are dimeric class C GPCRs that modulate neuronal excitability, synaptic plasticity and are drug targets for neurological disorders [117]. By smFRET, Vafabakhsh et al. probed the activation of full-length mammalian group II mGluR2 (Figure 3.8c) [42] and showed that the ligand-binding domains interconvert between three conformations: a resting, an activated and a short-lived intermediate state. Orthosteric agonists induce transitions between these conformational states with efficacy determined by occupancy of the active conformation. Their results supported a general mechanism for the activation of mGluRs in which agonist binding induces closure of the ligand-binding domains, followed by reorientation of the dimeric interface.

3.3.3 Biomolecular Interactions

SmFRET is also valuable in probing dynamic interactions between biomolecules, including binding mode transitions of the single-stranded DNA binding (SSB) protein on single-stranded DNA (ssDNA), nucleosome translocation by the ISWI (imitation switch)-family remodelers and phase separation induced by RNA–protein interactions.

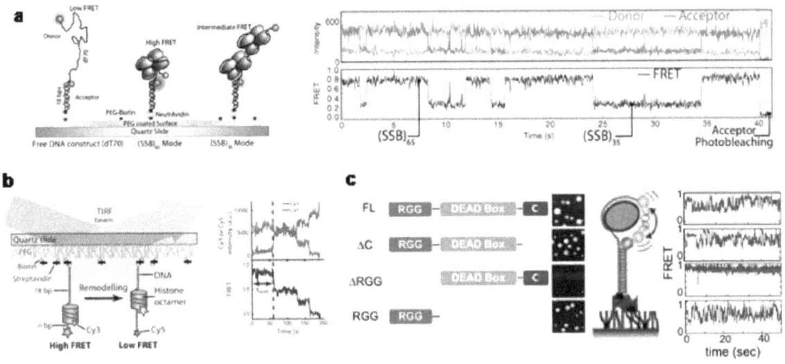

Figure 3.9 SmFRET assays of intermolecular interactions. **(a)** Conformational changes of ssDNA on SSB upon binding mode transitions. Reprinted with permission from ref. [83]. **(b)** DNA translocation on a nucleosome. Reprinted with permission from ref. [85]. **(c)** RGG domain plays a key role in phase separation of the disordered P granule protein LAF-1. Reprinted with permission from ref. [86].

SSB, which binds ssDNA in a sequence-independent manner, is essential in all organisms [118, 119]. It also modulates the functions of many DNA processive enzymes either via protein–protein interactions or by controlling its access to ssDNA [118–122]. Roy et al. applied smFRET to examine the dynamic interconversions among SSB/DNA complexes. They presented direct evidence of fluctuations between two major binding modes of SSB, $(SSB)_{35}$ and $(SSB)_{65}$ formed with $(dT)_{70}$ (Figure 3.9a) [83]. They further revealed a novel low abundance binding configuration and provided a direct demonstration that the SSB–ssDNA complexes are a finely tuned assembly in dynamic equilibrium among several well-defined structural and functional states.

The accessibility of DNA in nucleosomes and higher order chromatin structures is regulated by ATP-dependent chromatin-remodeling enzymes. Depending on the subunit composition, chromatin-remodelers can display divergent remodeling activities. For example, ISWI-family enzymes reposition nucleosomes while maintaining their canonical structure, whereas the SWI/SNF-family enzymes which translocate the nucleosome also change the nucleosome structure [123]. Using smFRET, Deindl et al. probed nucleosome translocations by the ISWI-family remodelers [85], and found that nucleosome remodeling begins with a 7 bp step of DNA translocation followed by 3 bp subsequent steps toward the exit side of nucleosomes (Figure 3.9b). In accordance with this, DNA movement on the entry side of the nucleosome occurs only after 7 bp of exit-side translocation, and each entry-side step draws in a 3 bp equivalent of DNA that allows three additional base pairs to be moved to the exit side. These results suggest a remodeling mechanism with well-defined coordination at different nucleosomal sites.

RNA granules, which are intracellular RNA/protein assemblies, function in diverse modes of RNA processing, including splicing, degradation and translational repression of mRNA [124]. Elbaum-Garfinkle et al. showed that the *Caenorhabditis elegant* protein LAF-1, a DDX3 RNA helicase found in P granules, phase-separates into P granule-like droplets in vitro [86]. SmFRET assays suggested that this RNA fluidization results from highly dynamic RNA–protein interactions that emerge at the droplet phase boundary. The authors demonstrated that an N-terminal intrinsically disordered protein domain of LAF-1 is necessary and sufficient for both phase separation and RNA–protein interactions (Figure 3.9c). These findings by smFRET demonstrated that LAF-1 is important for promotion of P granule assembly.

3.4 New Developments of smFRET

New strategies have been recently developed to overcome shortcomings of conventional smFRET. These include multicolor smFERT [37, 38], strategies to break the concentrations barrier [125, 126], smFRET under forces [19, 79, 127–129], SIFA [40], and quenchers-in-a-liposome FRET (lipoFRET) [41].

3.4.1 Multicolor smFRET

Conventional smFRET measures the energy transfer between a single FRET pair. However, the ability to observe distance changes between more than one FRET pair is desired when the analyte being studied shows a high degree of complexity. The FRET among three or more spectrally distinct fluorophores can provide a more complete picture of the conformation of biomolecules [38, 130–136]. While three-color smFRET measures three interfluorophore FRET efficiencies in real-time (donor and accepter$_1$, donor and accepter$_2$, accepter$_1$ and accepter$_2$), the four-color FRET technique can determine, in real-time six interfluorophore FRET efficiencies (donor and acceptor$_1$, donor and acceptor$_2$, donor and acceptor$_3$, acceptor$_1$ and acceptor$_2$, acceptor$_1$ and acceptor$_3$, acceptor$_2$ and acceptor$_3$) (Figure 3.10a). Monitoring all these FRET efficiencies can be achieved by alternating laser excitation (ALEX) with a switching-box triggered by EMCCD [38, 66].

In general, a three-color smFRET can reveal two correlated or uncorrelated conformational events. For example, the four-way Holliday junction is a simple model that undergoes two-state conformational fluctuations when only two fluorophores are labeled. By labeling three or four arms of the Holliday junction, distance changes between each donor–acceptor pair can be measured simultaneously and more complicated dynamics can be revealed (Figure 3.10b) [37]. Another example is the DNA dynamics in nucleosomes. Nucleosomes, which are a tightly packed form of DNA, are widely observed in eukaryotes and impede DNA transcription and replication. ATP-dependent chromatin remodeling factors catalyze nucleosome sliding to regulate gene transcription. With three-color smFRET, recent work showed that the movement of the entry-side DNA is earlier than that of the exit-side DNA during chromatin remodeling (Figure 3.10c), revealing a new mechanism of DNA translocation during chromatin remodeling [137].

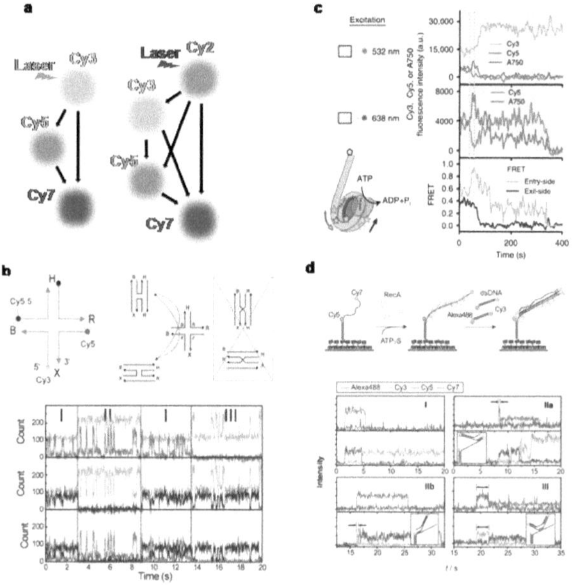

Figure 3.10 Multicolor smFRET. **(a)** Scheme of energy transfers in three- and four-color smFRET. **(b)** Three-color smFRET analysis of the dynamics of the four-way Holliday junction. Reprinted with permission from ref. [37]. **(c)** Coordinated DNA movements on a nucleosome during chromatin remodeling in the three-color smFRET. Reprinted with permission from ref. [137]. **(d)** The four-color smFRET provides more details on the RecA-mediated strand exchange in homologous recombination. Reprinted with permission from ref. [38].

Compared with three-color smFRET, four-color smFRET is capable of probing more than two events or two long-range molecular interactions [38]. For example, Lee et al. developed a "dual FRET pair" scheme, in which two independent FRET pairs measure the correlation between two spatially well-separated events and determine their synchronicity (Figure 3.10d). In this case, A FRET pair of Cy5/Cy7 labeled at two ends of 70-nt ssDNA were immobilized on a surface. Then, the RecA proteins were added and formed a RecA filament which displaces the homologous double-stranded DNA (dsDNA). Two donors (Cy2 and Cy3) were labeled on the two ends of the dsDNA. In this way, one can confirm which end is involved in the first step in the strand displacement process. With this system, they found three modes of DNA displacement, in which the initiation of strand exchange can occur from both ends of DNA or from the middle, which is not detected by two-color FRET (Figure 3.10d).

3.4.2 Strategies to Break Concentrations Barrier

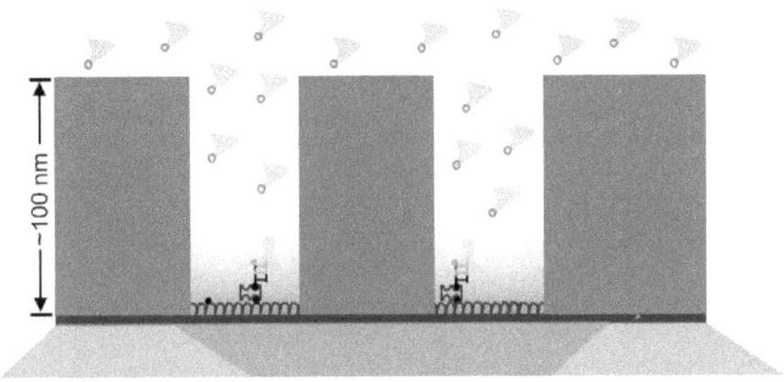

Figure 3.11 Signal-to-noise ratio enhancement in smFRET. Zero-mode waveguides generate a confined excitation field so that the laser excites only those biomolecules localized within the confined excitation volume, reducing the fluorescence background.

The concentrations of fluorophore-labeled freely-diffusing species are usually limited to a nanomolar concentration because of the lower signal-to-noise ratio and the high intensity of the background. However, observation of weak intermolecular interactions on a time scale accessible to conventional smFRET often requires very high concentrations of ligand biomolecules. This results in high background noise and consequently, a compromise of the quality of the smFRET signals. Three strategies have been developed to perform single-molecule measurements at high concentrations. The first of these uses self-assembled nano-antennas to enhance the signal [125]. In this configuration, one or two gold nanoparticles were attached to a DNA origami, which also incorporated docking sites for fluorophores next to one nanoparticle or in the gap between the two nanoparticles. The fluorescence can be enhanced by a factor of over 100 for a fluorophore positioned in the nano-gap between the gold nano-particles [125]. The second used zero-mode waveguides to reduce the observation volume and refine the interference of the background (Figure 3.11) [126], in which a confined excitation field was generated near the silica bottom of a nano-aperture, thereby selectively exciting only ligand biomolecules bound to a target biomolecule within a confined excitation volume [138]. The third method is the single-molecule photoactivation FRET (sm-PAFRET), a general approach to break the concentration barrier by using photoactivatable fluorophores as donors [39]. It was demonstrated that the smFRET time trace can be monitored in the presence of more than 1 μM acceptor-labeled proteins [126].

3.4.3 SmFRET under Forces

Forces may regulate biological processes. Many researchers expanded smFRET by combining it with force-based single-molecule manipulation methods, such as optical tweezers (Figure 3.12a) [127, 139], magnetic tweezers (Figure 3.12b) [79, 140, 141], atomic force microscope (Figure 3.12c) [128], DNA origami (Figure 3.12d) [129], or DNA nanotensioner (Figure 3.12e) [19]. These methods have advantages and shortcomings. For example, atomic force microscopes and optical tweezers in which only one molecule is measured at a time, are low in throughput. On the other hand, smFRET in combination with DNA origami and a DNA nanotensioner can measure hundreds of molecules in parallel.

Helicases, which are powered by nucleotide triphosphate hydrolysis, unwind dsDNA in discrete steps [142–145]. SmFRET has been widely applied to studies of helicases by detecting distance changes between a pair of dyes labeled on the overhangs of a forked DNA. However, it lacks the spatial resolution required to reveal stepping kinetics of the helicase. By bending a short DNA, Lin et al. designed a nanotensioner to exert a force on the overhangs (Figure 3.12e) [19]. The strategy improved the resolution of smFRET to 0.5 bp, and is capable of uncovering stepping of helicases during DNA unwinding. Ma et al. applied the method to characterization of the stepping of two nonring-shaped helicases, *Escherichia coli* (*E. coli*) RecQ and *Saccharomyces cerevisiae* Pif1 (ScPif1) [146]. They showed that when forked dsDNA with free overhangs (without the tension) were used as substrates, both helicases unwind the dsDNA in nonuniform steps that are widely distributed (Figure 3.13a). When a tension of ~6 pN was exerted on the overhangs, the overall profile of the step-size distribution of ScPif1 was narrowed, while that of *E. coli* RecQ remained unchanged (Figure 3.13b). The results indicated a different dependence of the helicase's stepping on force. The authors also proposed a universal stepping mechanism to interpret their data (Figure 3.13c).

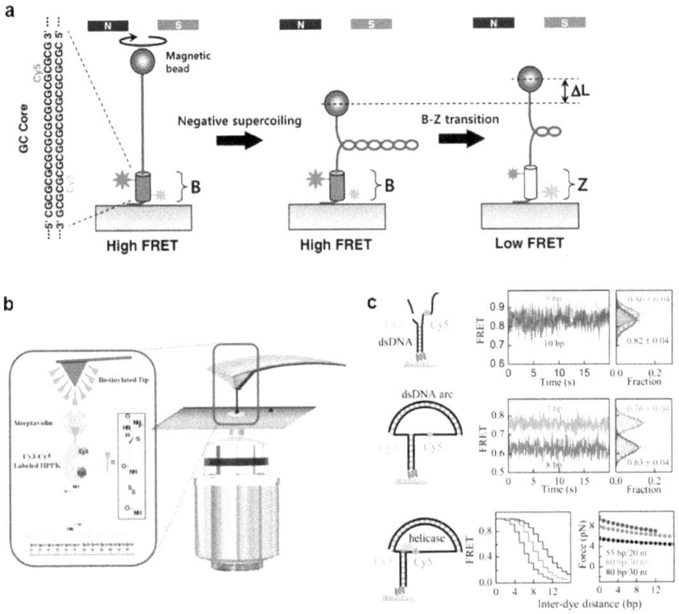

Figure 3.12 SmFRET combined with manipulation. **(a)** SmFRET with magnetic tweezers to study the B-Z transition under negative superhelicity. Reprinted with permission from ref. [79]. **(b)** SmFRET with atomic force microscope to study the protein conformational changes. Reprinted with permission from ref. [128]. **(c)** SmFRET with DNA nanotensioner to study the stepping of a molecular motor. Reprinted with permission from ref. [19].

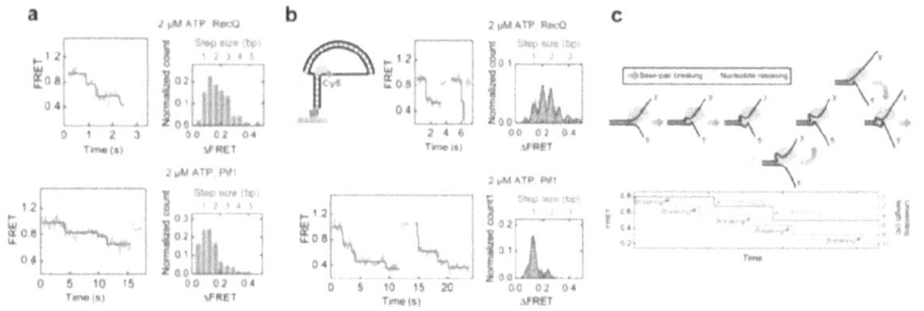

Figure 3.13 Nanotensioner-enhanced smFRET. Reprinted with permission from ref. [146]. **(a)** FRET with conventional forked DNA as the substrates of the RecQ and Pif1 helicases. **(b)** DNA nanotensioner used to improve the precision of FRET. **(c)** A universal model built to quantitatively explain the stepping of the helicases.

3.4.4 Surface-Induced Fluorescence Attenuation

Membrane proteins regulate various cellular activities and play vital roles in living cells. Research into the orientation and position of membrane proteins is crucial to an understanding of their functions. SmFRET has been used successfully to study membrane proteins. It works well when both the donor and the acceptor are prespecified. However, the donor and the acceptor might be too far apart to yield FRET signals when the proteins undergo diffusive motions. In most studies on membrane proteins, one is interested in just the position of the proteins in the reference frame of the membrane. The movement of a protein in a direction parallel to the membrane is usually concomitant with the movement perpendicular to the membrane because of the fluidity of the membrane. FRET however does not distinguish between these two kinds of movements. Li et al. recently developed a method known as SIFA to track both the vertical and lateral kinetics of singly labeled proteins in supported lipid bilayers (Figure 3.14) [40]. It is basically a point-to-plane distance indicator, and is well-suited to a study of the structural dynamics of membrane proteins. The quenching efficiency of SIFA follows the equation, $E = 1/\left[1+\left(d/d_0\right)^4\right]$, where d is the fluorophore-to-surface distance and d_0 is the characteristic distance at which half of the energy is transferred. For studies of proteins in lipid bilayers, a good choice of d_0 is approximately 4 nm (Figure 3.14). Li et al. employed graphene oxide (GO) as the quencher. If F and F_0 are the fluorescence in the presence and absence of the quencher respectively, the relative fluorescence of the donor, $F/F_0 = 1 - E$, decreases as the donor-labeled protein site approaches the quencher. In principle, SIFA can also provide information about the x–y coordinates of the protein trapped in the membrane simply by analyzing the images.

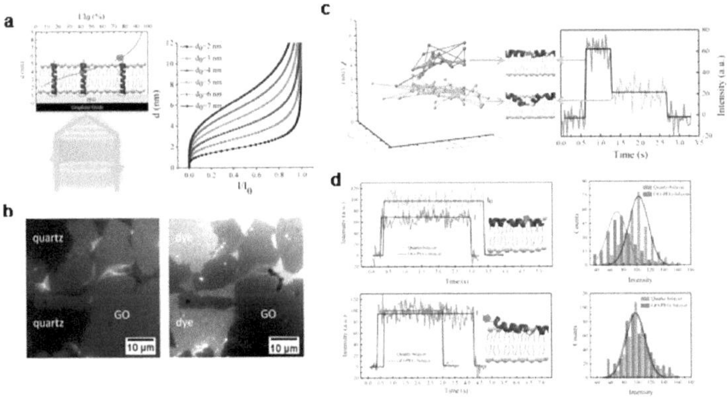

Figure 3.14 Principle of SIFA. Reprinted with permission from ref. [147].

3.4.5 Quenchers-in-a-Liposome FRET

SIFA is a good choice for studying protein dynamics on supported lipid bilayers. In many studies, a system which is more like a real, unsupported cell membrane may be desired. LipoFRET, a liposome-based method which takes advantage of the measurement principles of FRET, was developed to probe vertical position changes of a single protein on a membrane [41]. Application of the method to α-synuclein (α-syn) has distinguished the positions of specific sites relative to the membrane surface, demonstrating the high resolution and broad applicability of the method. LipoFRET is based on the FRET between one donor labeled on the membrane protein and multiple quenchers encapsulated in a liposome (Figure 3.15a). The energy transfer efficiency is given by $E = k_t/(t^{-1} + k_t)$, in which τ refers to the intrinsic lifetime of the donor, and k_t is the sum of pairwise energy transfer rates (Figure 3.15b). The relative fluorescence of the donor, $F/F_0 = 1 - E$, decreases as the donor-labeled protein site gets closer to the acceptors.

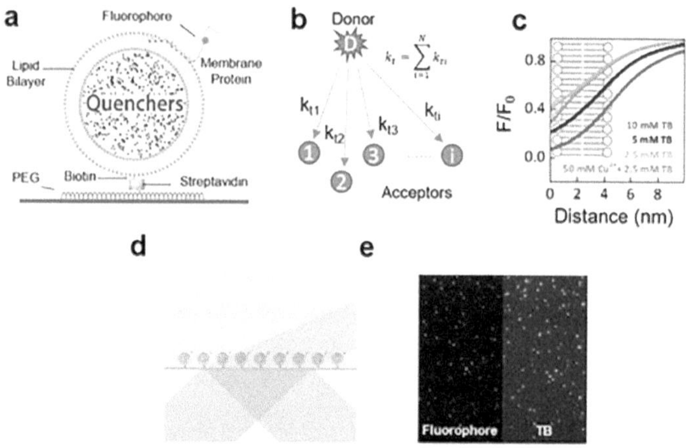

Figure 3.15 LipoFRET. Reprinted with permission from ref. [41]. **(a)** A fluorophore around a liposome full of quenchers. **(b)** The principle of pairwise energy transfers. **(c)** Calculated quenching efficiency against distance for various quencher concentrations used to convert fluorescence intensity into distance from the inner surface. **(d)** Schematic of the pseudo-TIRF illumination. **(e)** Typical images of fluorophores (α-syn T72C-Alexa555) and liposomes containing TB for the co-localization analysis.

During lipoFRET measurements, many dyes could be selected as the quenchers. The quenchers should not damage lipid membranes and must show an absorption spectrum that overlaps with the emission spectrum of the donor. The fluorescence of the quenchers should be weak enough to ensure minimal light leakage into the donor channel. For example, trypan blue (TB), which has been widely used to distinguish living and dead cells [148], can serve as the quencher. If one wants to detect dynamics outside the membrane, a higher concentration of quencher may be added to increase its sensitivity away from the membrane. On the other hand, the addition of Cu^{2+}-nitrilotriacetic acid complex (Cu-NTA) can further increase the sensitivity near the inner surface of the liposome. Monte Carlo simulation illustrates the relationship between F/F_0 and the distance of the donor to the inner surface of the liposome at different quencher concentrations (Figure 3.15c), in which TB and Cu-NTA were used. Such curves were used to convert intensities into distance with respect to the inner surface of the lipid bilayer. A standard two-channel FRET setup with pseudo-TIRF illumination (Figure 3.15d) can be used in lipoFRET experiments. A co-localization protocol can help to confirm that the quenching is indeed due to the quenchers in the liposome (Figure 3.15e). A beam splitter was used to separate signals of the donor and the quencher.

Application of lipoFRET to α-syn, which is critical in the pathology of Parkinson's disease and in presynaptic vesicle homeostasis, provided new knowledge about the dynamics of α-syn on membranes. Different membrane interaction patterns were observed with Alexa Fluor 555 (Alexa555) labeled on different sites of α-syn. For α-syn T72C-Alexa555 and α-syn S129C-Alexa555, the distribution of F/F_0 confirmed the labeled positions on the membrane surface and in the aqueous solution, respectively (Figure 3.16a and b). For Alexa555 labeled at the K10 site, the fluorescence transits slowly among multiple values in a time scale of a few seconds (Figure 3.16c). Three main peaks in the F/F_0 histogram correspond to the location at the outer surface of the lipid bilayer, 1.2 ± 0.7 nm below the outer surface and 3.4 ± 0.5 nm below the outer surface.

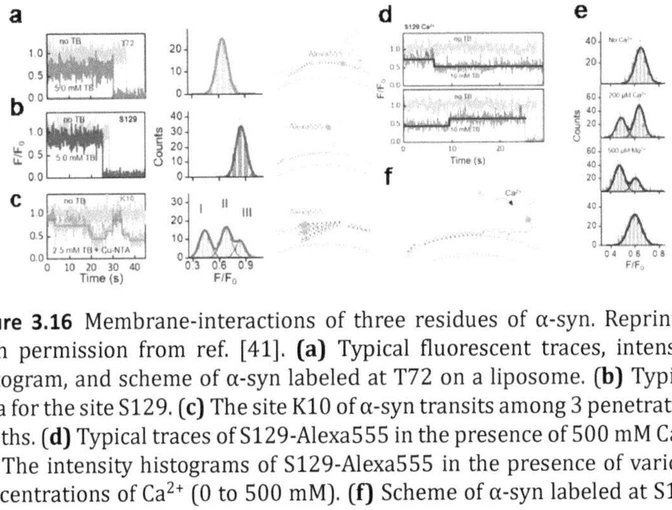

Figure 3.16 Membrane-interactions of three residues of α-syn. Reprinted with permission from ref. [41]. **(a)** Typical fluorescent traces, intensity histogram, and scheme of α-syn labeled at T72 on a liposome. **(b)** Typical data for the site S129. **(c)** The site K10 of α-syn transits among 3 penetration depths. **(d)** Typical traces of S129-Alexa555 in the presence of 500 mM Ca^{2+}. **(e)** The intensity histograms of S129-Alexa555 in the presence of various concentrations of Ca^{2+} (0 to 500 mM). **(f)** Scheme of α-syn labeled at S129 without or with Ca^{2+}.

Conformational changes of membrane proteins induced by ligands can also be detected with lipoFRET. Many researchers have proposed that the C-terminal tail of **α-syn is** probably involved in the Ca^{2+} binding process because α-syn is implicated functionally in signaling involving dopamine and Ca^{2+}. LipoFRET was used to examine the effect of Ca^{2+} on the position of S129 of the liposome-bound α-syn. A new peak corresponding to a lower position above the membrane arose in the F/F_0 histogram after the addition of Ca^{2+}. The fluorescence traces also showed transitions between two intensities (Figure 3.16d and e). The result showed that Ca^{2+} binding changed the C-terminus of α-syn to a new state in a concentration-dependent manner, with a position change of 1.2–1.6 nm (Figure 3.16f).

3.5 Summary and Perspective

In this chapter, we have reviewed fundamental concepts, technical implementations, and scientific applications of smFRET. Also recognized as a spectroscopic ruler, smFRET has evolved to be an indispensable tool for the investigation on a sub-nanometer scale of dynamic processes and transient conformational changes in heterogeneous systems. In order to overcome the shortcomings of conventional smFRET, many innovative strategies have been developed. Due mainly to space limitations, we have discussed only a few of them, including multi-color smFERT, strategies to break the concentrations barrier, smFRET under forces, SIFA and lipoFRET.

SmFRET has greatly enhanced our understandings of many biological processes. The next step would be the applications of smFRET in living cells. This would require intracellular labeling of biomolecules with small dyes. A variety of FRET probes were developed using fluorescent proteins and fusion-tags. However, such probes are typically large in size and this may limit their spatial resolution. Although much work remains, great efforts have been undertaken to label proteins inside living cells using external fluorophores [149]. Short peptide motifs have also been developed to bind covalently to synthetic organic fluorophores [62].

Appendix: A Tutorial Protocol: Single Molecule Dynamics of DNA G-Quadruplexes Measured by smFRET

Purpose

smFRET is capable of resolving the structural dynamics of individual biomolecules, such as DNA, RNA, or proteins. DNA G-quadruplex (G4), which is a specific DNA structure composed of four single strands of

DNA, was reported to have varying dynamic structures in different electrolyte buffer environments. Using G4 as a model analyte, the following tutorial protocol presents the core procedures of smFRET experiments, from preparing samples to data acquisition. Upon completion of the measurement, the readers will be able to directly visualize the structural dynamics of G4 and the regulation by salt.

Materials

- **Equipment**
  Inverted microscope (Olympus, IX71)
  Laser (Coherent, 532 nm)
  Dichroic mirror system (Cairn-research.co.uk, Optosplit II)
  EM-CCD (Andor, DU-897D-CS0-#BV)
  Optical table (Daeil, systems. Co., Ltd.)
  Pipettes (Eppendorf, 2.5 μL, 10μL, 50 μL and 200 μL)
  Autoclave (Sanyo Electric Co., Ltd. MLS-3750)
  Hot air oven (Venticell-MMM Medcenter Einrichtungen GmbH)
  Ultrasonic cleaner (BRANSON-3800)
  Drilling machine (MINIQ, BG-5166)
  Beaker (50–200 mL)
  Staining jar (50 mL)
- **Consumables**
  Pipette tips (Axygen, 20 μL, 200 μL. Autoclaved)
  Micro centrifuge tubes (Axygen, 200 μL. Autoclaved)
  Home-made reaction chamber (cover glass: Fisherbrand, 12-545-A. Figure 3.17a)
  Filter (Sartorius Minisart, 0.22 μm)
- **Reagents**
  Streptavidin (Sangon Biotech)
  G4 DNA strands labeled with Cy3 (5′-CGA CTG CTA GCA TGT CGT ACT AGC ACT GTA GCT GTA GTA CGG GTT AGG GTT AGG GTT AGG GT(T-Cy3) TTT TTT TTT TTT T-3′;) and Cy5 (5′-(C-Cy5)CC TAA CCC TAA CCC TAA CCC GTA CTA CAG CTA CAG TGC TAG TAC GAC ATG CTA GCA GTC G-biotin-3′) (Sangon Biotech)
  Tris hydrochloride (Tris-HCl, Sigma-Aldrich)
  Tris (Tris, Sigma-Aldrich)
  Sodium hydroxide (NaOH, Sigma-Aldrich)
  Sodium chloride (NaCl, Sigma-Aldrich)

Ultrapure water (ddH_2O, Milli-Q)
Photo protection system (including glucose oxidase (165 U/mL, –20°C), catalase (2170 U/mL, –20°C), β-D-glucose (0.4% wt/wt) and Trolox (–20°C) (Sigma-Aldrich)
Glacial acetic acid ($C_2H_4O_2$, Sigma-Aldrich)
(3-Aminopropyl) triethoxysilane (APTES, Sigma-Aldrich)
Sodium bicarbonate ($NaHCO_3$, Sigma-Aldrich)
Potassium sulfate (K_2SO_4, Sigma-Aldrich)
Acetone (C_3H_6O, MOS)
Methanol (CH_4O, MOS)
Ethanol (C_2H_6O, MOS)
Sulfuric acid (H_2SO_4, AR)
Hydrogen peroxide (H_2O_2, AR)
M-PEG-SVA (Laysan Bio, Inc. MW = 5,000)
Biotin--PEG-SVA (Laysan Bio, Inc. MW = 5,000)
Fluorescent microspheres (Invitrogen, F8786)

Safety and Precautions

NaOH is corrosive to metals and may cause skin corrosion and serious eye damage. It is exothermic when mixed with water. Many steps during washing of the cover glass or glass slide and the surface modification are dangerous. During these steps, wear appropriate gloves, lab coat, and face mask and handle materials in a fume hood.

Methods

- **Step 1: Solution Preparation**

  1.1 **Buffer preparation:** 20 mM Tris-HCl buffers (pH = 7.5 at 25°C) with different concentrations of NaCl (15 mM and 50 mM) are used. The buffers are autoclaved and membrane filtered. The prepared buffers can be stored at –20°C for long terms.

  1.2 **Oxygen-scavenging system:** The components of photo protection system are stored separately and mixed prior to each use. The solution of 50× β-D-glucose is prepared by dissolving β-D-glucose in ddH_2O with a 1:2.5 (w/v) ratio. 100× glucose oxidase (165 U/mL, –20°C) is obtained by mixing it with T50 buffer (20 mM Tris-HCl and 50 mM NaCl) in 1:10 (w/v). 100× catalase (2170 U/mL, –20°C) is obtained by mixing it with T50 buffer in 1:25 (w/v) ratio. The glucose oxidase and catalase solution can be stored for 1 or 2 weeks. 4× Trolox can be obtained by mixing Trolox with T50 buffer (pH 7.5) in 1:1,000 (w/v). The Trolox buffer can be kept at –20°C (keep in dark place) for long-term storage.

1.3 **Streptavidin solution:** Streptavidin (0.01 mg/mL in T50 buffer) is used to immobilize the biomolecules to the chamber surface.

1.4 **DNA solution:** To form G quadruplexes, DNA oligonucleotides (Materials) were dissolved in the annealing buffer (25 mM Tris-HCl, 100 mM NaCl, 1 mM EDTA), incubated at 95°C for 5 minutes, and then gradually cooled down to 25°C in 7 hours. The solution is then stored at –20°C prior to use.

- **Step 2: Setup Preparation**

2.1 **Simultaneous mapping of the donor and acceptor:** Photons from the donors and acceptors, which are respectively transferred to two different areas of an EMCCD camera (Figure 3.17b), should be paired to learn their transfer efficiency. This can be done with an image of fluorescent microspheres (~1 nM F8763) with substantial emission in both detection channels [24]. Record a single image of the fluorescent microspheres by the EM-CCD (Gain: 5, exposure time: 0.1 s). After manually selecting 3 to 4 fluorescence peaks in the donor image and their corresponding peaks in the acceptor image, an automated algorithm then calculates a linear transformation between the two images that corrects for offset, rotation, rescaling and distortion [24]. This linear transformation can then be used to map the donor and acceptor during subsequent measurements.

2.2 **Glass cleaning and modification:** The cover glass and the glass slide of the reaction chamber should be thoroughly cleaned because dust may emit fluorescence, causing interference to the measurement. (i) Place cover glasses and the glass slides in a staining jar; (ii) Add acetone in the staining jar and clean ultrasonically for 30 minutes. Wash the glass with ddH_2O by ultrasonic cleaning for 1 minute and repeat twice; (iii) Add methanol in the staining jar and ultrasonically clean for 30 minutes. Wash down the methanol with ddH_2O by ultrasonic cleaning for 1 minute and repeat twice;

(iv) Place another staining jar in a beaker and warm it to 95°C in a water bath; (v) Move the cover glasses and the glass slides into the second staining jar and keep the incubation at 95°C (water bath) for 120 minutes in a mixture of sulfuric acid and hydrogen peroxide with a volume ratio of 7:3 (also called the piranha solution). Cool the solution to room temperature. Wash down the piranha solution with ddH_2O by ultrasonic cleaning for 1 minute and repeat twice; (6) Place sodium ethoxide (a mixture of 2 g NaOH, 15 mL ddH_2O, 35 mL ethanol) in a staining jar and perform ultrasonic treatment for 15 minutes. Wash down the sodium ethoxide with ddH_2O in an ultrasonic cleaner for 15 minutes. Note: the above solutions are dangerous and should be operated with care.

Because biomolecules under investigation may adhere nonspecifically to the inner surface of the reaction chamber, the cover glass, and the glass slide should be passivated: (i) Place the cover glasses and the glass slide (in a staining jar) into the oven for 20 minutes (120°C) then cool them to room temperature; (ii) Add a mixture of 47.5 mL methanol, 2.5 mL glacial acetic acid, 0.5 mL APTES in the staining jar and set aside for 10 minutes. Wash the cover glasses and the glass slides with ddH_2O by ultrasonic cleaning for 5 minutes; (iii) Air-dry the cover glass and the glass slide with N_2; (iv) Add ~60 μL PEG coated system (30 mg m-PEG-SVA in 300 μL buffer, mix with 0.3 mg biotin-PEG-SVA in 3 μL buffer, dissolved prior to use with the buffer: 0.1 M $NaHCO_3$, 0.6 M K_2SO_4) in between two cover glasses or glass slides. Keep them in a petri dish surrounded with ddH_2O to maintain a humid environment for 150 minutes (avoid exposure of light); (v) Clean the glasses with ddH_2O and air dry them with N_2. Mark the side coated with PEG with a marker-pen.

2.3 **Reaction chamber preparation:** The reaction chamber is prepared as demonstrated in Figure 3.17a: (i) Drill 10 holes in the glass slide; (ii) Make five channels (~5 mm width) on a piece of double-sided tape; (iii) Cut a section of infusion tube (inner diameter about 2 mm) and join it to a 10 μL pipette tip as the outlet of the channel. The other end of the infusion tube should be attached to an injector; (iv) Use a 200 μL pipette tip as the inlet of the channel; (v) Assemble the reaction chamber.

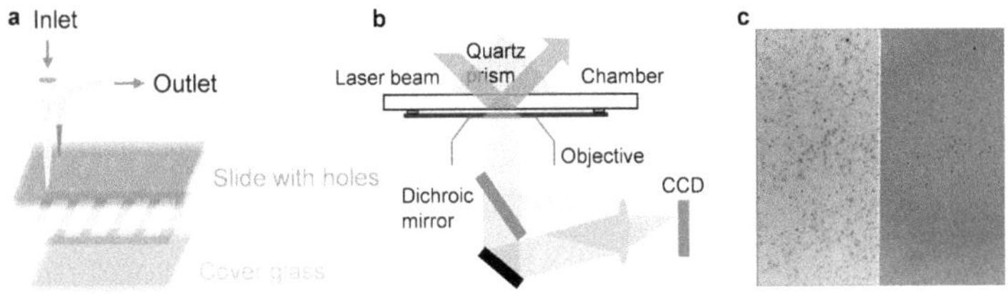

Figure 3.17 Experiment setup for smFRET. **(a)** Reaction chamber. **(b)** Photons from donors and accepters are separated and are respectively but simultaneously recorded in adjacent areas of the same EMCCD. **(c)** One frame of the film recorded by an EMCCD, hundreds of single-molecules in one view can be detected simultaneously.

2.4 **DNA immobilization:** About 100 pM (100 μL) biotin-labeled DNAs in a 200 μL pipette tip is jointed with one hole on the glass slide. The liquid is injected into the channel with a syringe and then incubated for 5 minutes, when the streptavidin has been immobilized on the inner chamber surface through streptavidin–biotin interaction (Figure 3.18a). Unbound DNA molecules are washed away with T50 buffer.

- **Step 3: Single-Molecule Sensing**

  Operating buffers with different NaCl concentrations are prepared by mixing the following buffers together: 71 μL Tris-HCl buffer containing 15 mM or 50 mM NaCl concentrations, 2 μL 50× β-D-glucose, 1 μL 100× glucose oxidase, 1 μL 100× catalase, 25 μL 4× Trolox. The mixed buffer solution is injected into the chamber to initiate the measurement. The sample is then ready to be imaged by the EM-CCD (Gain: 100, exposure time: 0.1 second) (Figure 3.17c), during which many single-molecule fluorescent spots of donors (Figure 3.17c, left) and acceptors (Figure 3.17c, right) can be observed, respectively. The imaging can be continuously carried out till most of the fluorescent spots become photo bleached.

- **Step 4: Data Analysis**

  To gain a quantitative understanding of the structure dynamics, the film can be quantitatively analyzed by software such as ImageJ. First, single-molecule florescence peaks should be found and paired. Then the local background around the molecule should be subtracted. Finally, one can obtain the donor and acceptor intensities for each molecule in a series of frames to gain the time traces of a pair of dyes (left panel in Figure 3.18b). Then the FRET time traces can be extracted (middle panel in Figure 3.18b). The hidden Markov chain model, enable one to fit the state transitions of the data (right panel in Figure 3.18b). At 50 mM NaCl, intermediate states [150] of a G4 DNA can be easily observed (bottom panel in Figure 3.18b). While for most traces in the 15 mM NaCl condition, the intermediate states of G4 are so unstable that only the fully folded state and fully unfolded state can be observed (top panel in Figure 3.18b).

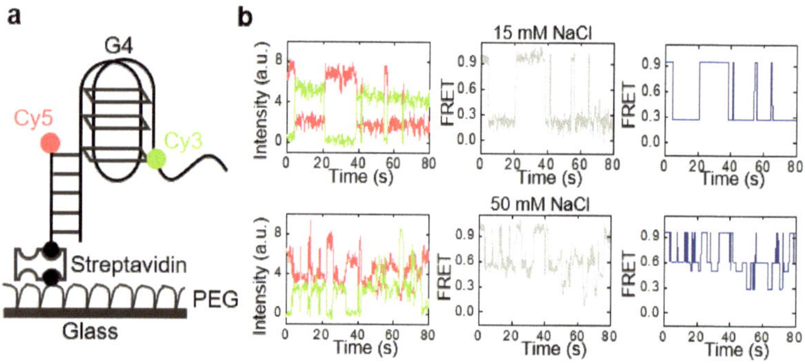

Figure 3.18 Dynamics of the G4 DNA. **(a)** A typical construction of G4 DNA for smFRET. **(b)** SmFRET traces and their states fitted by a hidden Markov model (blue lines) at 15 mM and 50 mM NaCl.

References

1. Franck J (1922). Einige aus der theorie von Klein und Bosseland zu ziehende folgerungen über fluoreszenz, photochemische prozesse und die elektronenemission glühender körper, *Z Phys*, **9**, doi:10.1007/bf01326976.

2. Carlo G (1922). Über Entstehung wahrer Lichtabsorption und scheinbare Koppelung von Quantensprüngen, *Z Phys*, **10**, doi:10.1007/bf01332559.

3. Cario G, Franck J (1922). Über zerlegung von Wasserstoffmolekülen durch angeregte Quecksilberatome, *Z Phys*, **11**, doi:10.1007/bf01328410.

4. Perrin J (1927). Fluorescence et induction moléculaire par résonance, *C R Acad Sci*, **184**, 1097–1100.

5. Lerner E, et al. (2018). Toward dynamic structural biology: two decades of single-molecule Forster resonance energy transfer, *Science,* **359**, doi:10.1126/science.aan1133.

6. Selvin PR (2000). The renaissance of fluorescence resonance energy transfer, *Nat Struct Biol,* **7**, 730–734.

7. Forster T (1965). *Modern Quantum Chemistry,* Sinanoglu O (ed), Academic, New York.

8. Stryer L, Haugland RP (1967). Energy transfer: a spectroscopic ruler, *Proc Natl Acad Sci U S A,* **58**, 719–726, doi:10.1073/pnas.58.2.719.

9. Arai Y, Nagai T (2013). Extensive use of FRET in biological imaging, *Microscopy (Oxf),* **62**, 419–428, doi:10.1093/jmicro/dft037.

10. Betzig E, Chichester RJ (1993). Single molecules observed by near-field scanning optical microscopy, *Science,* **262**, 1422–1425, doi:10.1126/science.262.5138.1422.

11. Shuming Nie DTC, Richard NZ (1994). Probing individual molecules with confocal fluorescence microscopy, *Science,* **266**, 4.

12. Betzig E, Trautman JK, Harris TD, Weiner JS, Kostelak RL (1991). Breaking the diffraction barrier: optical microscopy on a nanometric scale, *Science,* **251**, 1468–1470, doi:10.1126/science.251.5000.1468.

13. Betzig E, Trautman JK (1992). Near-field optics: microscopy, spectroscopy, and surface modification beyond the diffraction limit, *Science,* **257**, 189–195, doi:10.1126/science.257.5067.189.

14. Hirschfeld T (1976). Optical microscopic observation of single small molecules. *Applied optics* **15**, 2965-2966, doi:10.1364/AO.15.002965.

15. Moerner WE, Kador L (1989). Optical detection and spectroscopy of single molecules in a solid, *Phys Rev Lett,* **62**, 2535–2538, doi:10.1103/PhysRevLett.62.2535.

16. Orrit M, Bernard J (1990). Single pentacene molecules detected by fluorescence excitation in a p-terphenyl crystal, *Phys Rev Lett,* **65**, 2716–2719, doi:10.1103/PhysRevLett.65.2716.

17. Ha T. et al. (1996). Probing the interaction between two single molecules: fluorescence resonance energy transfer between a single donor and a single acceptor, *Proc Natl Acad Sci,* **93**.

18. Schuler B, Lipman EA, Steinbach PJ, Kumke M, Eaton WA (2005). Polyproline and the "spectroscopic ruler" revisited with single-molecule fluorescence, *Proc Natl Acad Sci U S A,* **102**, 2754–2759, doi:10.1073/pnas.0408164102.

19. Lin W, et al. (2017). Helicase stepping investigated with one-nucleotide resolution fluorescence resonance energy transfer, *Phys Rev Lett,* **119**, 138102, doi:10.1103/PhysRevLett.119.138102.

20. Choi UB, et al. (2010). Single-molecule FRET-derived model of the synaptotagmin 1-SNARE fusion complex, *Nat Struct Mol Biol,* **17**, 318–U384, doi:10.1038/nsmb.1763.

21. Kalinin S, et al. (2012). A toolkit and benchmark study for FRET-restrained high-precision structural modeling, *Nat Methods,* **9**, 1218–U1129, doi:10.1038/Nmeth.2222.

22. Hellenkamp B, Wortmann P, Kandzia F, Zacharias M, Hugel T (2017). Multidomain structure and correlated dynamics determined by self-consistent FRET networks, *Nat Methods,* **14**, 174–180, doi:10.1038/Nmeth.4081.

23. Eilert T, Beckers M, Drechsler F, Michaelis J (2017). Fast-NPS-A Markov Chain Monte Carlo-based analysis tool to obtain structural information from single-molecule FRET measurements, *Comput Phys Commun,* **219**, 377–389, doi:10.1016/j.cpc.2017.05.027.

24. Roy R, Hohng S, Ha T (2008). A practical guide to single-molecule FRET, *Nature Methods,* **5**, 507–516, doi:10.1038/nmeth.1208.

25. Ha T (2001). Single-molecule fluorescence resonance energy transfer, *Methods,* **25**, 78–86, doi:10.1006/meth.2001.1217.

26. Stracy M, Uphoff S, Garza de Leon F, Kapanidis AN (2014). In vivo single-molecule imaging of bacterial DNA replication, transcription, and repair, *FEBS Letters,* **588**, 3585–3594, doi:10.1016/j.febslet.2014.05.026.

27. Robinson A, van Oijen AM (2013). Bacterial replication, transcription and translation: mechanistic insights from single-molecule biochemical studies, *Nat Rev Microbiol,* **11**, 303–315, doi:10.1038/nrmicro2994.

28. Sustarsic M, Kapanidis AN (2015). Taking the ruler to the jungle: single-molecule FRET for understanding biomolecular structure and dynamics in live cells, *Curr Opin Struct Biol,* **34**, 52–59, doi:10.1016/j.sbi.2015.07.001.

29. Alhadid Y, et al. (2017). Studying transcription initiation by RNA polymerase with diffusion-based single-molecule fluorescence, *Protein Sci,* **26**, 1278–1290, doi:10.1002/pro.3160.

30. Munro JB, Vaiana A, Sanbonmatsu KY, Blanchard SC (2008). A new view of protein synthesis: mapping the free energy landscape of the ribosome using single-molecule FRET, *Biopolymers,* **89**, 565–577, doi:10.1002/bip.20961.

31. Lu HP (2011). Revealing time bunching effect in single-molecule enzyme conformational dynamics, *Phys Chem Chem Phys PCCP,* **13**, 6734–6749, doi:10.1039/c0cp02860f.

32. Weikl TR, Paul F (2014). Conformational selection in protein binding and function, *Protein Sci,* **23**, 1508–1518, doi:10.1002/pro.2539.

33. Lee SJ, Syed S, Ha T (2018). Single-molecule FRET analysis of replicative helicases, *Methods Mol Biol,* **1805**, 233–250, doi:10.1007/978-1-4939-8556-2_12.

34. Stryer L (1978). Fluorescence energy-transfer as a spectroscopic ruler, *Annu Rev Biochem,* **47**, 819–846, doi:10.1146/annurev.bi.47.070178.004131.

35. Lakowicz JR (1999). *Principles of Fluorescence Spectroscopyy,* 2nd ed, Plenum, New York.

36. Periasamy A (2001). Fluorescence resonance energy transfer microscopy: a mini review, *J Biomed Optics,* **6**, doi:10.1117/1.1383063].

37. Hohng S, Joo C, Ha T (2004). Single-molecule three-color FRET, *Biophys J,* **87**, 1328–1337, doi:10.1529/biophysj.104.043935.

38. Lee J, et al. (2010). Single-molecule four-color FRET, *Angew Chem,* **122**, 10118–10121, doi:10.1002/ange.201005402.

39. Peng S, Sun R, Wang W, Chen C (2017). Single-molecule photoactivation FRET: a general and easy-to-implement approach to break the concentration barrier, *Angew Chem Int Ed Engl,* **56**, 6882–6885, doi:10.1002/anie.201702731.

40. Li Y, et al. (2016). Single-molecule visualization of dynamic transitions of pore-forming peptides among multiple transmembrane positions, *Nat Commun,* **7**, 12906, doi:10.1038/ncomms12906.

41. Ma DF, et al. (2019). Detecting single-molecule dynamics on lipid membranes with quenchers-in-a-liposome FRET, *Angew Chem Int Ed Engl,* **58**, 5577–5581, doi:10.1002/anie.201813888.

42. Vafabakhsh R, Levitz J, Isacoff EY (2015). Conformational dynamics of a class C G-protein-coupled receptor, *Nature,* **524**, 497–501, doi:10.1038/nature14679.

43. Hinterdorfer P, Van Oijen A (2009). *Handbook of Single-Molecule Biophysics*, Springer.

44. Axelrod D (2013). Evanescent excitation and emission in fluorescence microscopy, *Biophys J,* **104**, 1401–1409, doi:10.1016/j.bpj.2013.02.044.

45. Zhang H, Guo P (2014). Single molecule photobleaching (SMPB) technology for counting of RNA, DNA, protein and other molecules in nanoparticles and biological complexes by TIRF instrumentation, *Methods,* **67**, 169–176, doi:10.1016/j.ymeth.2014.01.010.

46. Yang H, Xie XS (2002). Probing single-molecule dynamics photon by photon, *J Chem Phys,* **117**, 10965–10979, doi:10.1063/1.1521154.

47. Pirchi M, et al. (2016). Photon-by-photon hidden Markov model analysis for microsecond single-molecule FRET kinetics, *J Phys Chem B,* **120**, 13065–13075, doi:10.1021/acs.jpcb.6b10726.

48. Chung HS, Eaton WA (2018). Protein folding transition path times from single molecule FRET, *Curr Opin Struct Biol,* **48**, 30–39, doi:10.1016/j.sbi.2017.10.007.

49. Hartmann A, Berndt F, Ollmann S, Krainer G, Schlierf M (2018). In situ temperature monitoring in single-molecule FRET experiments, *J Chem Phys,* **148**, 123330, doi:10.1063/1.5008966.

50. Kim JY, Kim C, Lee NK (2015). Real-time submillisecond single-molecule FRET dynamics of freely diffusing molecules with liposome tethering, *Nat Commun,* **6**, 6992, doi:10.1038/ncomms7992.

51. Wang Z, Campos LA, Munoz V (2016). Single-molecule fluorescence studies of fast protein folding, *Methods Enzymol,* **581**, 417–459, doi:10.1016/bs.mie.2016.09.011.

52. Kapanidis AN, Weiss S (2002). Fluorescent probes and bioconjugation chemistries for single-molecule fluorescence analysis of biomolecules, *J Chem Phys,* **117**, 10953–10964, doi:10.1063/1.1521158.

53. Ha T, et al. (2002). Initiation and re-initiation of DNA unwinding by the *Escherichia coli* Rep helicase, *Nature,* **419**, 638–641, doi:10.1038/nature01083.

54. Sean A, McKinney ADJF, David MJL, Taekjip H (2004). Observing spontaneous branch migration of Holliday junctions one step at a time, *PNAS.*

55. Lee JY, Kim DS, Taekjip H (2009). Extreme conformational diversity in human telomeric DNA, *PNAS.*

56. Vafabakhsh R, Ha T (2012). Extreme bendability of DNA less than 100 base pairs long revealed by single-molecule cyclization, *Science,* **337**, 1097–1101, doi:10.1126/science.1224139.

57. Li JH, et al. (2016). Pif1 is a force-regulated helicase, *Nucleic Acids Res,* **44**, 4330–4339, doi:10.1093/nar/gkw295.

58. Diez M, et al. (2004). Proton-powered subunit rotation in single membrane-bound F0F1-ATP synthase, *Nat Struct Mol Biol,* **11**, 135–141, doi:10.1038/nsmb718.

59. Nikic I, Kang JH, Girona GE, Aramburu IV, Lemke EA (2015). Labeling proteins on live mammalian cells using click chemistry, *Nat Protocols*, **10**, 780–791, doi:10.1038/nprot.2015.045.

60. Leisle L, et al. (2016). Cellular encoding of Cy dyes for single-molecule imaging, *eLife*, **5**, doi:ARTN e1908810.7554/eLife.19088.

61. DeRocco V, Anderson T, Piehler J, Erie DA, Weninger K (2010). Four-color single-molecule fluorescence with noncovalent dye labeling to monitor dynamic multimolecular complexes, *BioTechn*, **49**, 807–816, doi:10.2144/000113551.

62. Hoffmann C, et al. (2005). A FlAsH-based FRET approach to determine G protein-coupled receptor activation in living cells, *Nat Methods*, **2**, 171–176, doi:10.1038/nmeth742.

63. Fernandes DD, et al. (2017). Characterization of Fluorescein Arsenical Hairpin (FlAsH) as a probe for single-molecule fluorescence spectroscopy, *Sci Rep*, **7**, 13063, doi:10.1038/s41598-017-13427-8.

64. Kapanidis AN, et al. (2004). Fluorescence-aided molecule sorting: analysis of structure and interactions by alternating-laser excitation of single molecules, *Proc Natl Acad Sci U S A*, **101**, 8936–8941, doi:10.1073/pnas.0401690101.

65. Uphoff S, et al. (2010). Monitoring multiple distances within a single molecule using switchable FRET, *Nat Methods*, **7**, 831–836, doi:10.1038/nmeth.1502.

66. Hohlbein J, Craggs TD, Cordes T (2014). Alternating-laser excitation: single-molecule FRET and beyond, *Chem Soc Rev*, **43**, 1156–1171, doi:10.1039/c3cs60233h.

67. Rasnik I, Myong S, Cheng W, Lohman TM, Ha T (2004). DNA-binding orientation and domain conformation of the *E. coli* Rep helicase monomer bound to a partial duplex junction: single-molecule studies of fluorescently labeled enzymes, *J Mol Biol*, **336**, 395–408, doi:10.1016/j.jmb.2003.12.031.

68. Jeyachandran YL, Mielczarski JA, Mielczarski E, Rai B (2010). Efficiency of blocking of non-specific interaction of different proteins by BSA adsorbed on hydrophobic and hydrophilic surfaces, *J Colloid Interface Sci*, **341**, 136–142, doi:10.1016/j.jcis.2009.09.007.

69. Hua B, et al. (2014). An improved surface passivation method for single-molecule studies, *Nat Methods*, doi:10.1038/nmeth.3143.

70. Prime KL, Whitesides GM (1991). Self-assembled organic monolayers - model systems for studying adsorption of proteins at surfaces, *Science*, **252**, 1164–1167, doi:10.1126/science.252.5009.1164.

71. Wang S, et al. (2015). Unwinding forward and sliding back: an intermittent unwinding mode of the BLM helicase, *Nucleic Acids Res,* **43**, 3736–3746.

72. Park J, et al. (2010). PcrA helicase dismantles RecA filaments by reeling in DNA in uniform steps, *Cell,* **142**, 544–555, doi:10.1016/j.cell.2010.07.016.

73. Arslan S, Khafizov R, Thomas CD, Chemla YR, Ha T (2015). Engineering of a superhelicase through conformational control, *Science,* **348**, 344–347, doi:10.1126/science.aaa0445.

74. Campos LA, et al. (2011). A photoprotection strategy for microsecond-resolution single-molecule fluorescence spectroscopy, *Nat Methods,* **8**, 143–146, doi:10.1038/nmeth.1553.

75. Rasnik I, McKinney SA, Ha T (2006). Nonblinking and long-lasting single-molecule fluorescence imaging, *Nat Methods,* **3**, 891–893, doi:10.1038/nmeth934.

76. Aitken CE, Marshall RA, Puglisi JD (2008). An oxygen scavenging system for improvement of dye stability in single-molecule fluorescence experiments, *Biophys J,* **94**, 1826–1835, doi:10.1529/biophysj.107.117689.

77. Deniz AA, et al. (2000). Single-molecule protein folding: diffusion fluorescence resonance energy transfer studies of the denaturation of chymotrypsin inhibitor 2, *Proc Natl Acad Sci U S A,* **97**, 5179–5184, doi:10.1073/pnas.090104997.

78. Hellenkamp B, et al. (2018). Precision and accuracy of single-molecule FRET measurements—a multi-laboratory benchmark study, *Nat Methods,* **15**, 669–676, doi:10.1038/s41592-018-0085-0.

79. Lee M, Kim SH, Hong SC (2010). Minute negative superhelicity is sufficient to induce the B-Z transition in the presence of low tension, *Proc Natl Acad Sci U S A,* **107**, 4985–4990, doi:10.1073/pnas.0911528107.

80. Xiaowei Zhuang HK, Pereira MJB, Babcock HP, Walter NG, Chu S (2002). Correlating structural dynamics and function in single ribozyme molecules, *Science.*

81. David Rueda, GB, Rhodes MM, Rust MJ, Zhuang X, Walter NG (2004). Single-molecule enzymology of RNA: essential functional groups impact catalysis from a distance, *PNAS.*

82. Muller-Spath S, et al. (2010). From the cover: charge interactions can dominate the dimensions of intrinsically disordered proteins, *Proc Natl Acad Sci U S A,* **107**, 14609–14614, doi:10.1073/pnas.1001743107.

83. Roy R, Kozlov AG, Lohman TM, Ha T (2007). Dynamic structural rearrangements between DNA binding modes of *E. coli* SSB protein, *J Mol Biol,* **369**, 1244–1257, doi:10.1016/j.jmb.2007.03.079.

84. Yodh JG, Stevens BC, Kanagaraj R, Janscak P, Ha T (2009). BLM helicase measures DNA unwound before switching strands and hRPA promotes unwinding reinitiation, *EMBO J,* **28**, 405–416, doi:10.1038/emboj.2008.298.

85. Deindl S, et al. (2013). ISWI remodelers slide nucleosomes with coordinated multi-base-pair entry steps and single-base-pair exit steps, *Cell,* **152**, 442–452, doi:10.1016/j.cell.2012.12.040.

86. Elbaum-Garfinkle S, et al. (2015). The disordered P granule protein LAF-1 drives phase separation into droplets with tunable viscosity and dynamics, *Proc Natl Acad Sci USA,* **112**, 7189–7194.

87. Lisby M, Rothstein R (2004). DNA damage checkpoint and repair centers, *Curr Opin Cell Biol,* **16**, 328–334, doi:10.1016/j.ceb.2004.03.011.

88. Kowalczykowski SC (2015). An overview of the molecular mechanisms of recombinational DNA repair, *Cold Spring Harb Perspect Biol,* **7**, doi:10.1101/cshperspect.a016410.

89. Polleys EJ, House NCM, Freudenreich CH (2017). Role of recombination and replication fork restart in repeat instability, *DNA Repair,* **56**, 156–165, doi:10.1016/j.dnarep.2017.06.018.

90. Lilley DMJ (2017). Holliday junction-resolving enzymes-structures and mechanisms, *FEBS Lett,* **591**, 1073–1082, doi:10.1002/1873-3468.12529.

91. Gibbs DR, Dhakal S (2019). Homologous recombination under the single-molecule fluorescence microscope, *Int J Mol Sci,* **20**, doi:10.3390/ijms20236102.

92. Hurley LH (2001). Secondary DNA structures as molecular targets for cancer therapeutics, *Biochem Soc Trans,* **29**, 692–696, doi:10.1042/0300-5127:0290692.

93. Kolesnikova S, Curtis EA (2019). Structure and function of multimeric G-quadruplexes, *Molecules,* **24**, doi:10.3390/molecules24173074.

94. Holliday R (1964). Mechanism for gene conversion in fungi, *Genet Res,* **5**, 282, doi:10.1017/S0016672300001233.

95. Potter H, Dressler D (1978). In vitro system from *Escherichia coli* that catalyzes generalized genetic-recombination, *Proc Natl Acad Sci USA,* **75**, 3698–3702, doi:10.1073/pnas.75.8.3698.

96. Schwacha A, Kleckner N (1995). Identification of double holliday junctions as intermediates in meiotic recombination, *Cell,* **83**, 783–791, doi:10.1016/0092-8674(95)90191-4.

97. Sen D, Gilbert W (1988). Formation of parallel 4-stranded complexes by guanine-rich motifs in DNA and its implications for meiosis, *Nature,* **334**, 364–366, doi:10.1038/334364a0.

98. Zahler AM, Williamson JR, Cech TR, Prescott DM (1991). Inhibition of telomerase by G-quartet DNA structures, *Nature,* **350**, 718–720, doi:10.1038/350718a0.

99. Lei M, Podell ER, Cech TR (2004). Structure of human POT1 bound to telomeric single-stranded DNA provides a model for chromosome end-protection, *Nat Struct Mol Biol,* **11**, 1223–1229, doi:10.1038/nsmb867.

100. Zaug AJ, Podell ER, Cech TR (2005). Human POT1 disrupts telomeric G-quadruplexes allowing telomerase extension in vitro, *Proc Natl Acad Sci U S A,* **102**, 10864–10869, doi:10.1073/pnas.0504744102.

101. Sun H, Karow JK, Hickson ID, Maizels N (1998). The Bloom's syndrome helicase unwinds G4 DNA, *J Biol Chem,* **273**, 27587–27592, doi:10.1074/jbc.273.42.27587.

102. Fry M, Loeb LA (1999). Human Werner syndrome DNA helicase unwinds tetrahelical structures of the fragile X syndrome repeat sequence d(CGG)(n), *J Biol Chem,* **274**, 12797–12802, doi:10.1074/jbc.274.18.12797.

103. Schaffitzel C, et al. (2001). In vitro generated antibodies specific for telomeric guanine-quadruplex DNA react with Stylonychia lemnae macronuclei, *Proc Natl Acad Sci U S A,* **98**, 8572–8577, doi:10.1073/pnas.141229498.

104. Blackwood EM, Kadonaga JT (1998). Going the distance: a current view of enhancer action, *Science,* **281**, 60–63, doi:10.1126/science.281.5373.60.

105. Wong OK, Guthold M, Erie DA, Gelles J (2008). Interconvertible lac repressor-DNA loops revealed by single-molecule experiments, *PLoS Biol,* **6**, 2028–2042, doi:ARTN e23210.1371/journal.pbio.0060232.

106. Richmond TJ, Davey CA (2003). The structure of DNA in the nucleosome core, *Nature,* **423**, 145–150, doi:10.1038/nature01595.

107. Baumann CG, Smith SB, Bloomfield VA, Bustamante C (1997). Ionic effects on the elasticity of single DNA molecules, *Proc Natl Acad Sci U S A,* **94**, 6185–6190.

108. Gesteland RF, Cech T, Atkins JF (1999). *The RNA World : The Nature of Modern RNA Suggests a Prebiotic RNA*, 2nd ed, Cold Spring Harbor Laboratory Press.

109. Treiber DK, Williamson JR (2001). Beyond kinetic traps in RNA folding, *Curr Opin Struct Biol,* **11**, 309–314, doi:10.1016/S0959-440x(00)00206-2.

110. Walter NG, Harris DA, Pereira MJB, Rueda D (2001). In the fluorescent spotlight: global and local conformational changes of small catalytic RNAs, *Biopolymers,* **61**, 224–242.

111. Thirumalai D, Lee N, Woodson SA, Klimov DK (2001). Early events in RNA folding, *Annu Rev Phys Chem,* **52**, 751–762, doi:10.1146/annurev.physchem.52.1.751.

112. Moore PB, Steitz TA (2003). The structural basis of large ribosomal subunit function, *Annu Rev Biochem,* **72**, 813–850, doi:10.1146/annurev.biochem.72.110601.135450.

113. Silverman SK (2003). Rube Goldberg goes (ribo)nuclear? Molecular switches and sensors made from RNA, *RNA,* **9**, 377–383, doi:10.1261/rna.2200903.

114. Haas E (2005). The study of protein folding and dynamics by determination of intramolecular distance distributions and their fluctuations using ensemble and single-molecule FRET measurements, *Chemphyschem,* **6**, 858–870, doi:10.1002/cphc.200400617.

115. Makhatadze GI, Privalov PL (1995). Energetics of protein structure, *Adv Protein Chem, Vol 47,* **47**, 307–425, doi:10.1016/S0065-3233(08)60548-3.

116. Uversky VN, Oldfield CJ, Dunker AK (2008). Intrinsically disordered proteins in human diseases: introducing the D-2 concept, *Ann Rev Biophys,* **37**, 215–246, doi:10.1146/annurev.biophys.37.032807.125924.

117. Niswender CM, Conn PJ (2010). Metabotropic glutamate receptors: physiology, pharmacology, and disease, *Annu Rev Pharmacol Toxicol,* **50**, 295–322, doi:10.1146/annurev.pharmtox.011008.145533.

118. Chase JW, Williams KR (1986). Single-stranded-DNA binding-proteins required for DNA-replication, *Annu Rev Biochem,* **55**, 103–136, doi:10.1146/annurev.biochem.55.1.103.

119. Meyer RR, Laine PS (1990). The single-stranded DNA-binding protein of *Escherichia coli, Microbiol Rev,* **54**, 342–380.

120. Yuzhakov A, Kelman Z, O'Donnell M (1999). Trading places on DNA-a three-point switch underlies primer handoff from primase to the replicative DNA polymerase, *Cell,* **96**, 153–163, doi:10.1016/S0092-8674(00)80968-X.

121. Sun WL, Godson GN (1998). Structure of the *Escherichia coli* primase/single-strand DNA-binding protein/phage G4ori(c) complex required for primer RNA synthesis, *J Mol Biol,* **276**, 689–703, doi:10.1006/jmbi.1997.1471.

122. Lohman TM, Ferrari ME (1994). *Escherichia coli* single-stranded DNA-binding protein—multiple DNA-binding modes and cooperativities, *Annu Rev Biochem,* **63**, 527–570, doi:10.1146/annurev.bi.63.070194.002523.

123. Clapier CR, Cairns BR (2009). The biology of chromatin remodeling complexes, *Annu Rev Biochem,* **78**, 273–304, doi:10.1146/annurev.biochem.77.062706.153223.

124. Anderson P, Kedersha N (2006). RNA granules, *J Cell Biol,* **172**, 803–808, doi:10.1083/jcb.200512082.

125. Acuna GP, Holzmeister P, Beater S, Lalkens B, Tinnefeld P (2012). Fluorescence enhancement at docking sites of DNA-directed self-assembled nanoantennas, *Science.*

126. Colin D. Kinz-Thompson, MP, Pulukkunat DK, Chenet D, Hone J, Wind SJ, Gonzalez RL, Jr (2013). Robustly passivated, gold nanoaperture arrays for single-molecule fluorescence microscopy, *ACS Nano.*

127. Hohng S, et al. (2007). Fluorescence-force spectroscopy maps two-dimensional reaction landscape of the holliday junction, *Science,* **318**, 279–283, doi:10.1126/science.1146113.

128. He Y, Lu M, Cao J, Lu HP (2012). Manipulating protein conformations by single-molecule AFM-FRET nanoscopy, *ACS Nano,* **6**, 1221–1229, doi:10.1021/nn2038669.

129. Nickels PC, Phil Holzmeister BW, Bae W, et al. (2016). Molecular force spectroscopy witha DNA origami–based nanoscopicforce clamp, *Science.*

130. Lee NK, Koh HR, Han KY, Kim SK (2007). Folding of 8-17 deoxyribozyme studied by three-color alternating-laser excitation of single molecules, *J Am Chem Soc,* **129**, 15526–15534, doi:10.1021/ja0725145.

131. Roy R, Kozlov AG, Lohman TM, Ha T (2009). SSB protein diffusion on single-stranded DNA stimulates RecA filament formation, *Nature,* **461**, 1092–1097, doi:10.1038/nature08442.

132. Munro JB, et al. (2010). Spontaneous formation of the unlocked state of the ribosome is a multistep process, *Proc Natl Acad Sci U S A,* **107**, 709–714, doi:10.1073/pnas.0908597107.

133. Lee NK, Koh HR, Han KY, Lee J, Kim SK (2010). Single-molecule, real-time measurement of enzyme kinetics by alternating-laser excitation fluorescence resonance energy transfer, *Chem Commun,* **46**, 4683–4685, doi:10.1039/c002666b.

134. Shin S, et al. (2016). Active control of repetitive structural transitions between replication Forks and Holliday junctions by werner syndrome helicase, *Structure,* **24**, 1292–1300, doi:10.1016/j.str.2016.06.004.

135. Ratzke C, Hellenkamp B, Hugel T (2014). Four-color FRET reveals directionality in the Hsp90 multicomponent machinery, *Nat Commun,* **5**, 4192, doi:10.1038/ncomms5192.

136. Chung HS, et al. (2017). Oligomerization of the tetramerization domain of p53 probed by two- and three-color single-molecule FRET, *Proc Natl Acad Sci USA,* **114**, E6812–E6821, doi:10.1073/pnas.1700357114.

137. Sabantsev A, Levendosky RF, Zhuang XW, Bowman GD, Deindl S (2019). Direct observation of coordinated DNA movements on the nucleosome during chromatin remodelling, *Nat Commun,* **10**, doi:ARTN 172010.1038/s41467-019-09657-1.

138. Levene MJ, et al. (2003). Zero-mode waveguides for single-molecule analysis at high concentrations, *Science,* **299**, 682–686, doi:10.1126/science.1079700.

139. Comstock MJ, et al. (2015). Direct observation of structure-function relationship in a nucleic acid–processing enzyme, *Science,* **348**, 352–354.

140. Long X, Parks JW, Bagshaw CR, Stone MD (2013). Mechanical unfolding of human telomere G-quadruplex DNA probed by integrated fluorescence and magnetic tweezers spectroscopy, *Nucleic Acids Res,* **41**, 2746–2755, doi:10.1093/nar/gks1341.

141. Kemmerich FE, et al. (2016). Simultaneous single-molecule force and fluorescence sampling of DNA nanostructure conformations using magnetic tweezers, *Nano Lett,* **16**, 381–386, doi:10.1021/acs.nanolett.5b03956.

142. Patel SS, Picha KM (2000). Structure and function of hexameric helicases, *Annu Rev Biochem,* **69**, 651–697, doi:10.1146/annurev.biochem.69.1.651.

143. Singleton MR, Dillingham MS, Wigley DB (2007). Structure and mechanism of helicases and nucleic acid translocases, *Annu Rev Biochem*, **76**, 23–50, doi:10.1146/annurev.biochem.76.052305.115300.

144. Lohman TM, Tomko EJ, Wu CG (2008). Non-hexameric DNA helicases and translocases: mechanisms and regulation, *Nat Rev Mol Cell Biol*, **9**, 391–401, doi:10.1038/nrm2394.

145. von Hippel PH (2004). Helicases become mechanistically simpler and functionally more complex, *Nat Struct Mol Biol*, **11**, 494–496, doi:10.1038/nsmb0604-494.

146. Ma JB, et al. (2018). Asynchrony of base-pair breaking and nucleotide releasing of helicases in DNA unwinding, *J Phys Chem B*, **122**, 5790–5796, doi:10.1021/acs.jpcb.8b01470.

147. Ma L, Li Y, Ma J, Hu S, Li M (2018). Watching three-dimensional movements of single membrane proteins in lipid bilayers, *Biochemistry*, **57**, 4735–4740, doi:10.1021/acs.biochem.8b00253.

148. Srivastava GK, et al. (2011). Trypan blue staining method for quenching the autofluorescence of RPE cells for improving protein expression analysis, *Exp Eye Res*, **93**, 956–962, doi:10.1016/j.exer.2011.07.002.

149. Teng KW, et al. (2016). Labeling proteins inside living cells using external fluorophores for microscopy, *eLife*, **5**, doi:10.7554/eLife.20378.

150. Hou XM, et al. (2015). Molecular mechanism of G-quadruplex unwinding helicase: sequential and repetitive unfolding of G-quadruplex by Pif1 helicase, *Biochem J*, **466**, 189–199, doi:10.1042/BJ20140997.

Chapter 4

DNA Origami as Single-Molecule Biosensors

Travis A. Meyer,[a] Qinyi Lu,[b] Kristin Weiss,[a] and Yonggang Ke[a,b]
[a]*Wallace H. Coulter Department of Biomedical Engineering, Emory University and Georgia Institute of Technology, Georgia, USA*
[b]*Department of Chemistry, Emory University, Georgia, USA*
yke2@emory.edu

Since the invention of the DNA origami technology in 2006, there have been a considerable amount of experimental studies utilizing this platform for the detection and analysis of single molecules. DNA origami nanostructures, constructed with global addressability and nanometer precision through bottom-up self-assembly, uniquely position the technology to facilitate a wide variety of single-molecule studies that are not feasible with other techniques. In this chapter, most of the published studies using DNA origami for single-molecule sensing and analysis are highlighted. These studies include the use of DNA origami nanostructures to detect the presence of protein and nucleic acid targets, as well as the study of biomolecular interactions, chemical reactions, and photonic techniques. A discussion of the challenges associated with use of DNA origami for single-molecule

Single-Molecule Tools for Bioanalysis
Edited by Shuo Huang

ISBN 978-981-4800-44-0 (Hardcover), 978-1-003-18913-8 (eBook)
www.jennystanford.com

biosensing and potential future avenues for improvement are included. Finally, a brief protocol for the design, synthesis, and characterization of a simple 2D DNA origami structure is provided.

4.1 Introduction

Over the past three decades, the information-carrying biomolecule DNA has seen an ever-expanding role as a material for the construction of nanoscale objects [1]. The field of structural DNA nanotechnology began in the 1980s, when Nadrian Seeman proposed rigid DNA motifs based on branched structures seen in Holliday Junctions as a platform for arrangement of proteins in crystallographic experiments [2]. DNA has several useful characteristics for a construction material; it is a chemically stable biomolecule with a well-defined structure and appealing mechanical properties. Furthermore, the highly specific molecular recognition capability of DNA, mediated by Watson–Crick base-pairing, enables the programmed self-assembly of a large number of unique DNA sequences into well-defined higher-order architectures. The relatively simplistic nature of DNA, both from structural and recognition perspectives, makes de novo design of DNA nanostructures more feasible than more complex protein assemblies.

DNA nanostructures have gradually increased in complexity since that initial hypothesis, from a simple cube with double helical edges [3] to massive, arbitrarily prescribed shapes formed from over 400 unique oligonucleotides [4]. In particular, the invention of the DNA origami technique in 2006 by Paul Rothemund was a quantum leap in the capacity for nanofabrication by enabling researchers to design and assemble arbitrarily-shaped complex three-dimensional objects [5]. This technique offers a number of key advantages when compared to other structural DNA technologies, including higher yields of correctly formed structures and higher stability of the final product. DNA origami involves the use of a long single-stranded DNA (ssDNA) "scaffold" strand, which is "folded" into a user-defined architecture using hundreds of short, synthetic "staple" strands, which are designed to be complementary to different domains of the scaffold strand (Figure 4.1A). The staple strands crosslink spatially distant regions of the scaffold, inducing the formation of parallel bundles of DNA double helices held together by strands crossing from one helix to another. By carefully designing the sequence-complementarity between the scaffold strand and the mixture of staple strands, the scaffold strand can be folded into a wide-variety of user-defined architectures. Several excellent reviews provide a more detailed elaboration of the DNA origami design principles [6, 7].

Following the initial development of Rothemund's two-dimensional (2D) rectangle, subsequent work focused on creation of more complex designs, such as the folding of 2D sheets into hollow, three-dimensional (3D) objects [8, 9]. Another landmark improvement came in 2009 when William Shih's group reported on the synthesis of solid 3D objects based on organizing DNA helices into both honeycomb and square lattices (Figure 4.1B) [10, 11]. At the same time, researchers from the same group developed strategies to induce curvature into DNA origami structures by selectively adding and removing bases from the staple strands [12]. Further advances in DNA origami design have led to the design of intricate wire-frame patterns [13–15]. Due to size limitations placed on traditional DNA origami due to the use of a scaffold strand, larger DNA supramolecular structures have been formed using hierarchical assemblies of multiple DNA origami subunits [16–19].

The design and synthesis of DNA origami structures typically begins with the use of computational programs such as caDNAno, which allow researchers to graphically plot the position and dimensions of DNA double helices which will make up the final structure [20]. Users are then able to choose which scaffold strand to use, and the software outputs sequences of all staple strands needed for the design. Following the synthesis of these staple strands, typically by a commercial producer, the structures can be folded by annealing a mixture of scaffold strand, staple strands, and the appropriate buffer. Divalent cations such as magnesium are traditionally included in the annealing mixture in order to minimize electrostatic repulsion between closely packed negatively charged helices. The Protocols section at the end of this chapter has a more in-depth look at the design and synthesis of DNA origami structures.

Advances in both the design and synthesis of complex DNA origami nanostructures have led to a large increase in their use for practical applications [21–25]. In particular, DNA origami can be used for nanofabrication of materials other than nucleic acids, particularly to direct the arrangement of these materials in precise positions and orientations. These interactions can be engineered through the use of chemically modified staple strands (e.g., biotin, thiols, amines, alkynes, azides, or digoxin) or through the use of "capture" strands, which consist of staples with ssDNA extensions protruding from the surface that can hybridize to complementary sequences conjugated to a species of interest, such as a protein or a nanoparticle. By selectively designing staple strands that are either chemically modified or contain "capture" extensions, the position of other molecular species can be organized (Figure 4.1C).

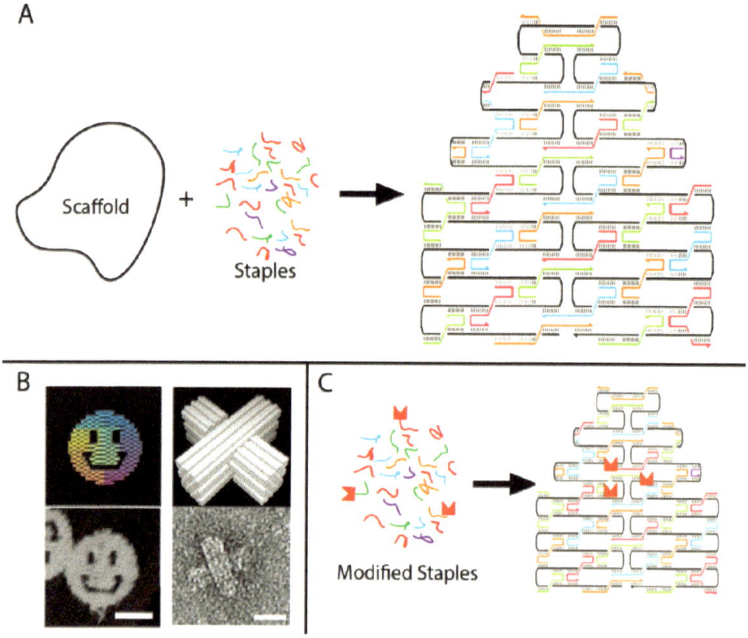

Figure 4.1 Summary of DNA origami. (A) DNA origami is formed by directing the folding of a long scaffold strand through hybridization with a large number of short synthetic staple strands. Figure adapted with permission from ref. [5]. Copyright 2006, Nature Publishing Group. (B) Examples of schematics (top) and microscopy images (bottom) of two-dimensional and three-dimensional DNA origami nanostructures. Reproduced with permission from ref. [5]. Copyright 2006, Nature Publishing Group, [10]. Copyright 2009, Nature Publishing Group. (C) Introduction of modifications to select staple strands can be used to add functionality/sensing elements to specific sites on the final DNA origami nanostructure.

The use of DNA origami as a platform for the detection and analysis of single-molecules was first reported in 2008 [26] and has since become a major focus in the field. DNA origami nanostructures have been used to detect specific genes and single nucleotide polymorphisms (SNPs) [27, 28], to monitor the action of enzymes at the single molecule level [29, 30], and even to study the transfer of energy in complex pathways [31]. DNA origami has numerous significant advantages that make it a particularly appealing technology for these single-molecule studies; (1) the high precision and global addressability, (2) the large and stable profile, and (3) the ease of bottom-up self-assembly.

1. Global Addressability and Nanoscale Precision: Because DNA origami is constructed from a unique scaffold, the relative position of each base is known within a few nanometers. This level of precision and addressability is one of the hallmarks of DNA origami and has particular significance for single-molecule biosensing, enabling improvements in detection limits through the placement of high-density sensing elements as well as the capacity for multiplexed assaying by easy identification of heterogeneous sensing elements within an asymmetric nanostructure.
2. Large and Stable Profile: The large size of DNA origami nanostructures helps to facilitate identification and localization of single molecules, particularly with techniques such as atomic force microscopy (AFM) or transmission electron microscopy (TEM). The large surface area of the nanostructures also provides more stable interactions with substrates due to their larger surface areas, which helps for dynamic imaging in which the same molecule is visualized over a period of time.
3. Ease of Bottom-Up Self Assembly: The self-assembly of DNA origami structures enables the simultaneous production of large numbers of single-molecule sensors. This enables these materials to be produced more easily and cheaply than other single-molecule sensors which require top-down synthetic schemes, such as electron-beam lithography, which requires a large amount of time and effort.

For the sake of organization, we have classified studies using DNA origami for single-molecule biosensing into four different categories. In Section 4.2, we discuss the use of DNA origami for the detection of both protein and nucleic acid targets. In Section 4.3, we expand the scope of the single-molecule detection studies to focus on the study of the activity and interactions between biomolecules, which includes binding between molecules, protein activity, and the formation of alternative nucleic acid conformations. In Section 4.4, we discuss how DNA origami nanostructures are used to control and visualize chemical reactions at the single-molecule level, while Section 4.5 discusses examples of studying photonic processes such as Forster resonance energy transfer (FRET) and surface-enhanced Raman spectroscopy (SERS) with applications in biotechnology.

4.2 Protein and Nucleic Acid Detection

Single-molecule biosensing has a significant advantage over traditional sensing approaches, namely ultra-low detection limits. Additionally, the minimal volume requirements allow for the detection of samples in which either the samples or the reagents are rare, expensive, or difficult to obtain. The detection of proteins and sequence-specific DNA molecules is of critical interest within the clinical and biomedical fields as a means of effectively diagnosing and monitoring specific illnesses or the outcomes of treatment, and consequently significant research effort has been devoted to developing DNA origami nanostructures capable of single-molecule detection of these two species.

There are two primary techniques for visualizing the presence of a target analyte using DNA origami biosensors. This first involves direct visualization of target molecules upon binding to the DNA origami structure, typically through the use of AFM or TEM. In this instance, binding of the target molecule leads to a change in the mechanical properties or height profile of the DNA nanostructure, which can then be observed by changes in the AFM signal. This strategy is relatively easy to incorporate and allows for comparatively simple DNA origami structures, but the types of species which can be detected using this technique is more limited. The second technique involves the coupling of binding between sensing elements with large-scale reconfigurations of the DNA nanostructure. These large-scale conformational changes can be easier to observe using techniques such as AFM and TEM, which makes analysis of the binding events more straightforward. Furthermore, this strategy also enables the use of alternative measurement techniques beyond AFM and TEM, such as optical techniques including fluorescence measurements via the FRET-effect or optical tweezers. Finally, the coupling of binding to large-scale conformational changes can be accomplished using target species which would normally be undetectable based on AFM imaging of static binding, such as small molecules and ions. However, this technique necessitates more complex nanostructures which are more difficult to both design and synthesize.

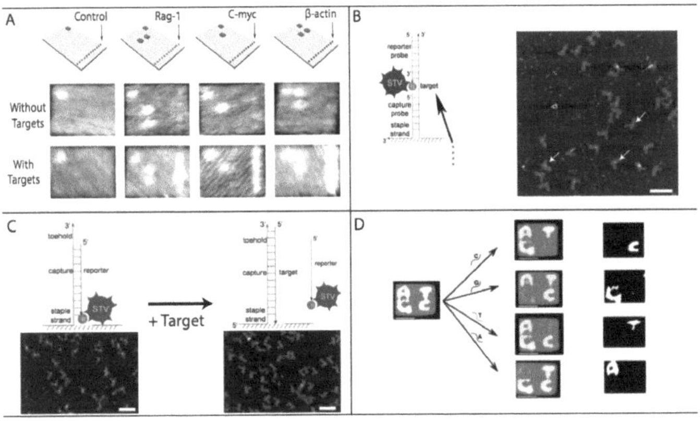

Figure 4.2 Direct visualization of target binding. (A) Binding of select mRNA targets to probes on DNA origami enables visualization and detection using AFM. Dumbbell-shaped protrusions are used to barcode individual tiles for identification of a target. Reproduced with permission from ref. [26]. Copyright 2008, AAAS. (B) Binding of streptavidin to biotin-labeled reporter probes in presence of target enables identification of target on index-free asymmetric "Map of China" DNA origami structure. Reproduced with permission from ref. [32]. Copyright 2010, Wiley-VCH. (C) Single-nucleotide polymorphism detection is enabled by toe-hold mediated strand displacement of biotinylated reporter probe with target DNA sequence. Reproduced with permission from ref. [36]. Copyright 2010, Wiley-VCH. (D) Single-nucleotide polymorphism detected using strand-displacement reaction of dumbbell-shaped dsDNA probes on surface of rectangular DNA origami leading to the disappearance of probes making letter of individual SNP. Reproduced with permission from ref. [27]. Copyright 2011, American Chemical Society.

For strategies based on direct visualization of binding, the pioneering work in this area was accomplished by Ke et al., in which authors demonstrated the first use of the DNA origami technique as a platform for the detection of specific nucleic acids [26]. The group designed a two dimensional (2D) rectangular tile with V-shaped probes complementary to specific genes extending from the surface. Upon hybridization of the probe with its complementary target, differences in the mechanical properties of single-stranded (free probes) and double-stranded (occupied probes) DNA enable AFM visualization of the bound target (Figure 4.2A). The design also included dumbbell-shaped protrusions on one corner of the DNA origami tile, which act as an index to allow unambiguous determination of the tile orientation and thus the probe location and identity. Incubation with target mRNA led to the appearance of clear "bulges" in AFM images, while nonspecific binding with sequence mismatch strands or background cellular RNA had no effect. Multiplexed assaying was accomplished using a mixture of target-specific tiles with additional dumbbell protrusions arranged in unique patterns to enable tile, and thus target, identification. Since hybridization efficiency increased linearly with increasing [target]:[probe] ratios, the authors were able to make semiquantitative measurements on the level of target present in the sample, although this required ab initio estimates of target concentration to optimize probe ratios. The authors were able to achieve 2- to 20-fold increases in maximum sensitivity compared surface-based probes like those used in DNA microarrays. However, the authors hypothesize that, with improvements in liquid handling and sufficient dilution of tiles to allow for nanoliter volumes of picomolar concentrations to be deposited on the surface, as few as 1,000 copies of target sequences could be detected. This detection limit would theoretically allow for the analysis of single-cell gene expression.

In a similar work, Zhang et al. used asymmetric DNA origami tiles for the single-molecule detection of target DNA molecules (Figure 4.2B) [32]. The group used a "Map of China" DNA origami design that was established previously, because the highly asymmetric nature allows the entire surface to be unambiguously mapped without the need for additional indexing such as the dumbbell-shaped protrusions used in the work by Ke et al. [26]. This index-free methodology potentially opens up the entire surface for probe placement, increasing the capacity for multiplex assays. Rather than using the differences in mechanical properties of double-stranded and ssDNA to distinguish between bound and unbound probes, the authors used biotinylated DNA sequences and streptavidin binding to generate AFM contrast. This strategy creates a more distinctive signal change, but the need for biotinylated target sequences renders this strategy infeasible for real world applications. To address this limitation, the authors used a sandwich based assay, in which the probe sequence was split between two different DNA strands—one half as an extension of the original staple strand, and the other half a biotinylated sequence that was free in solution. Only upon target binding by both halves is the biotinylated strand brought to the DNA nanostructure surface, where streptavidin binding and subsequent AFM imaging could detect the presence of the target sequence. While the use of linear probes and index-free design do provide the added benefit of increasing the multiplexing capabilities of the system, the use of biotinylated sequences and thus the added cost is a significant disadvantage.

Building on of these pioneering studies, similar designs have been used for the detection of new targets by other groups, including HPV genes isolated from human subjects [28], malaria protein biomarkers using a matching aptamer [33], and the quantitation of microRNA [34]. Small molecules such as aflatoxin B1 have also been imaged using aptamers and competitive gold nanoparticle (AuNP) binding as an AFM read-out [35].

In 2010, Zhang et al. extended the functionality of the "Map of China"-shaped DNA origami nanosensors to detect single nucleotide polymorphisms (SNPs) (Figure 4.2C) [36].The identification of SNPs has garnered significant interest recently due to the role of these mutations in various genetic diseases. However, the specificity needed to distinguish between targets with such minor differences has proven to be a considerable challenge, especially in equilibrium conditions where the thermodynamic differences between target and mismatch strands are often too small to allow for sufficient discrimination. In order to overcome this limitation, the authors developed a kinetically controlled toe-hold mediated strand displacement reaction from the surface of the DNA origami nanochip. This currently well-established technique involves a target strand binding to a single-stranded region on a template duplex (the toe-hold) and displacing the incumbent strand. Once bound to the toe-hold, the target displaces the incumbent strand through branch migration. If the target strand contains one or more mismatched oligonucleotides relative to the template strand, the branch migration process stalls, and the strand displacement does not reach completion. In the study, the authors annealed a biotinylated incumbent sequence to select staple strands containing single-stranded overhangs complementary to the target sequence, along with a 4-nt toe-hold region. The addition of the perfectly complimentary target sequence displaced the biotinylated strand, leading to a loss of streptavidin binding at that location, while the majority of streptavidin labels remained when single-base mismatched controls were used.

In a report published in Nano Letters in 2011, Subramanian et al. also developed a platform for the use of kinetically controlled branch migration processes to differentiate between SNPs on a DNA origami structures (Figure 4.2D) [27]. The design consisted of a 2D rectangular DNA origami nanostructure with four different types of strand-displacement probes extending from select staple strands. Each probe contained a SNP at a single site corresponding to the four possible bases (A, T, G, and C) and were organized in such a way as to produce graphical representations of the characters A, T, G, and C, with each probe consisting of a V-shaped dsDNA protrusion (similar to that in Ke et al. [26]) with an accessible toe-hold. Following addition of the target sequence, only the probes fully complementary to the target sequence were displaced, leaving ssDNA staple overhangs in those positions and thus generating a negative AFM signal. While the unambiguous graphical readout and use of dsDNA signal generation rather than STV binding are benefits compared to the previous work, the authors failed to report on the sensitivity of the system, using target concentrations in the micromolar range, and the ability to create multiplexed assays is limited.

For the strategy involving the coupling of target binding to large-scale conformational changes in the DNA origami structure, the first DNA origami design utilizing this strategy was presented in a report by Kuzya for the detection of DNA and other molecular species, such as metal ions and proteins (Figure 4.3A) [37]. The authors designed a DNA origami nanodevice, termed "DNA origami pliers," which consists of two 6 helix-bundle lever domains connected at the fulcrum by a

Holiday junction. This hinge can have three different conformations: an antiparallel right-handed X shape, an antiparallel linear shape, and a parallel linear shape, with the X-shape the most energetically favorable conformation. The authors designed a small concavity in each arm for binding to the target molecule—these concavities lay side-by-side in the parallel linear conformation to form a single large cavity. The DNA origami nanostructure was designed to exhibit three different nanomechanical shape transitions between the different conformations, which could be visualized using AFM: pinching, zipping, and unzipping.

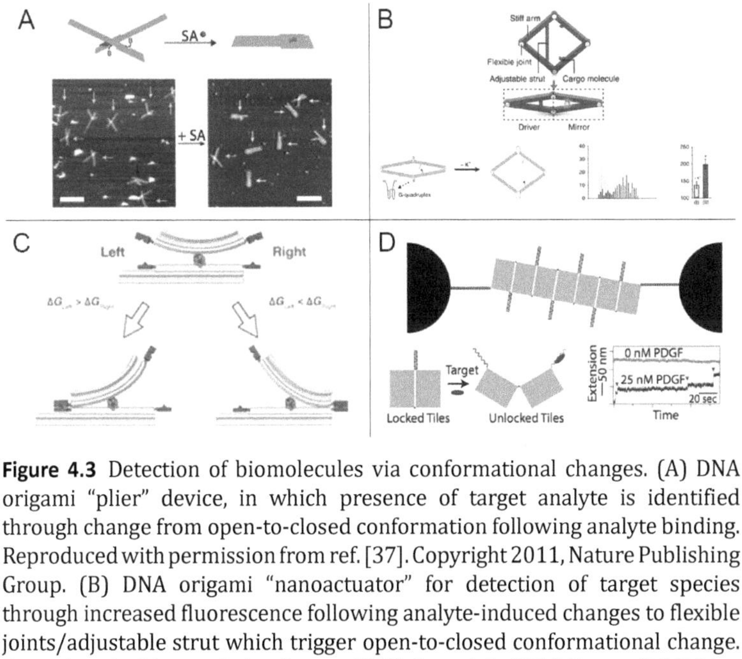

Figure 4.3 Detection of biomolecules via conformational changes. (A) DNA origami "plier" device, in which presence of target analyte is identified through change from open-to-closed conformation following analyte binding. Reproduced with permission from ref. [37]. Copyright 2011, Nature Publishing Group. (B) DNA origami "nanoactuator" for detection of target species through increased fluorescence following analyte-induced changes to flexible joints/adjustable strut which trigger open-to-closed conformational change. Reproduced with permission from ref. [39]. Copyright 2016, Nature Publishing Group. (C) Seesaw DNA origami device capable of detecting differences in strength of biomolecular interactions via competitive conformational change. Reproduced with permission from ref. [40]. Copyright 2016, Wiley-VCH. (D) Nanomechanical DNA origami probe which detects target binding via change in mechanical properties of DNA origami chain and subsequent shift in optical tweezer position. Reproduced with permission from ref. [41]. Copyright 2014, Wiley-VCH.

All three strategies were based on use of the presence of target analytes to induce changes between different conformations, which could be visualized using AFM or monitored in solution with fluorescence microscopy. The pinching motion involves the binding of a single-target molecule to ligands located in both concavities on the arms, bringing the DNA origami structures from the X-shaped conformation to a parallel linear conformation, and was demonstrated to detect model targets streptavidin binding and anti-fluorescein antibody binding using biotin- and fluorescein-modified DNA probes. However, a major limitation to this motion is that target molecules must be multivalent in order to bind to ligands in both concavities. The zipping motion was utilized to detect target molecules in which interactions with ligands were too weak to facilitate closing of the DNA origami pliers with a single molecule. For this design, the authors included multiple binding sites along the length of the lever arms, such that cooperative binding of all ligands could bring the device into the parallel linear conformation. Sensing of both potassium and sodium ions using G-quadruplex sensing motifs was demonstrated as well as silver ions using C–C mismatch accommodation. The unzipping technique involves the opposite transformation, in which the removal of preformed zipping elements could be used to detect target molecules through the transition from parallel linear to X-shaped conformation. Detection of target DNA sequences was accomplished using toe-hold mediated strand displacement of preformed zipping duplexes, while ATP detection was accomplished using ATP-specific aptamers as the zipping element. Changing the shape of the plier structures facilitated multiplexed assaying, while the use of two different locking duplexes in the same structure was used to enable AND gate logic operations. In a subsequent study, the authors integrated both the pinching and zipping method into a single structure, which facilitated allosteric control of sensing behavior and improved binding yields [38].

The use of a large DNA origami nanomechanical device as a transducer to turn molecular binding events into large-scale mechanical motions facilitated the detection of molecular species over a wide range of sizes, from metal ions to large proteins with only small changes to the design of the DNA origami structure. While the initial range of possible target molecules was limited, the ability to theoretically "invert" the pinching mechanism to include the attachment of antibodies into the concavities potentially opens up this technique to any antigen. While capable of the multiplexed assaying through the use of structures with unique geometries comparable to other techniques, the ability to include AND logic gates into the zipping and unzipping motions adds an interesting additional functionality—the ability to detect the presence of specific combinations of RNA/DNA species thought to be involved in certain pathologies is especially noteworthy.

Utilizing large-scale conformational changes as a read-out for the detection of molecular species has also been utilized by other groups. Ke et al. used a conformational change in a trapezoidal "nanoactuator" from a prestressed, compressed form to an open form to detect the presence of molecular species capable of disrupting the locking duplexes; specific DNA sequences, potassium ions, and restrictions enzymes were used as model targets (Figure 4.3B) [39]. Liedl et al. developed another similar system in which comparative binding events led to different conformations of a DNA origami "seesaw" (Figure 4.3C) [40]. Using a strand-displacement reaction initiated by a target DNA sequence, the seesaw could change conformations that were detectable using TEM.

Large-scale conformational changes induced by target–element interactions can also be detected with other techniques, such as optical tweezers. Optical tweezers enable the application of small forces (picoNewtons) and measurement of small movements (nanometers) by trapping micron-sized glass beads using highly focused laser beams. While this technique has some significant advantages over other detection schemes, such as very high signal-to-noise ratios, the throughput is a major hindrance—current strategies using optical tweezers only allow for a single measurement to be made at a time.

In work by Koirala et al., the authors used a DNA origami nanostructure as a platform to combine multiple sensing elements into a single entity, thus increasing the throughput of the biosensor (Figure 4.3D) [41]. The authors designed a DNA origami device which consists of seven 2D rectangular tiles linked sequentially on one corner by routing of the scaffold strand. The opposite corner contains ssDNA extensions, such that hybridization between complementary extensions on neighboring tiles "locks" the tiles in an edge-to-edge conformation rather than the "unlocked" corner-to-corner conformation. Transition from the locked to the unlocked state through dehybridizaton of the locking extension leads to a significant increase in flexibility of the structure, which can be measured using the optical tweezers. For the proof-of-principle studies, the authors were able to detect platelet-derived growth factor (PDGF) using locking strands derived from a PDGF-aptamer. Binding of the aptamer to PDGF released the locking strands and could be detected as a displacement change in the optical tweezers with a detection limit of around 10 pM. Multiplexed detection was also demonstrated using two different locking sequences, the PDGF aptamer and a toe-hold-containing duplex complementary to a target sequence. The benefits of this "mechanochemical" optical tweezer detection system include small background signals, high signal-to-noise ratios, and improved detection limits through the incorporation of multiple recognition sites which improves the detection limit. However, the use of this system is probably limited to studies on mechanoanalytic chemistry rather than wholescale biosensing, as the limited throughput and access to optical tweezer systems detracts from its usefulness for traditional biosensing assays. Furthermore, it is unclear if this design could ever be used to provide quantitative data on the amount of target species present rather than a binary present-not present signal, though it is potentially feasible that a time to unlocking event could be correlated with target concentration.

In conclusion, both direct binding and conformational changes have been successfully utilized to detect the presence of both DNA and protein analytes at the single-molecule level. Future efforts should be focused on expanding this technology to a wider variety of clinically relevant targets, as well as further development of detection techniques more amenable to the simple, high-throughput set-up required for processing clinical samples.

4.3 Analysis of Biomolecular Interactions and Activity

While DNA origami has been used to demonstrate the ability for single-molecule biosensing, the technique has proven extraordinarily successful in the study of various processes at the single-molecule level. In these instances, it is not the presence of a specific target that is of interest, but rather how that target interacts with either the DNA origami structure itself or other target molecules. These kinds of studies are critical to an improved understanding of a wide variety of biological and chemical processes, as they provide a direct observation of the process. These kinds of studies also facilitate the understanding of heterogeneous processes or the identification of outliers, both of which can be lost in bulk, ensemble measurements.

DNA origami nanostructures not only provide a scaffold to improve observation of the targets using probe measurements such as AFM, but also enable the precise spatial organization and design constraints to ask more complex scientific questions. In this section, we will discuss examples of the use of DNA origami nanostructures for this single-molecule analysis in several different areas: the analysis of molecular interactions, the analysis of protein activity, the analysis of alternative nucleic acid structures, the analysis of chemical reactions and finally, the analysis of photonic behavior.

Molecular interactions in which two species bind reversibly play a ubiquitous role in all aspects of biology—protein–protein, protein–ligand, DNA-binding proteins, DNA hybridization are all vitally important to proper cellular function, and dysfunction of these processes frequently leads to disease states. Furthermore, molecular interactions have been extensively used in biotechnology and biomedical applications, ranging from the development of specific diagnostic techniques to the development of novel pharmacological compounds. While ensemble measurements can provide a tremendous amount of information about the wide range of molecular interactions, single-molecule studies on these processes can be used to reveal the heterogeneous behavior of subpopulations that are hidden in bulk analysis. Furthermore, single-molecule studies on molecular interactions can be used validate proposed hypotheses on interaction mechanisms as well as generate new models based on information that can be gained only from direct analysis/visualization of the binding reactions.

Outside of the traditional benefits afforded to single-molecule studies using DNA origami, the ability to impose restraints on DNA double-helices further allows for researchers to conduct mechanistic studies on how DNA topology (bending, etc.) effects molecular interactions. While advancements in the functionalization of oligonucleotides with diverse chemical groups has opened up the ability to study a wider range of molecular interactions, the nature of the DNA origami simplifies the study of molecular interactions involving nucleic acids, including DNA-binding proteins and aptamer-protein complexes. As such, the majority of current studies have focused on these kinds of interactions.

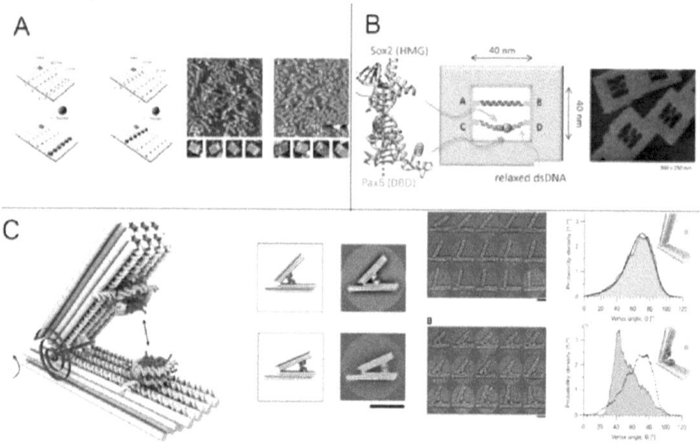

Figure 4.4 Single-molecule analysis of biomolecular interactions. (A) Binding of thrombin to aptamer pairs arranged on DNA origami nanostructure is controlled by the distance between the two DNA aptamers. Reproduced with permission from ref. [42]. Copyright 2008, Nature Publishing Group. (B) Preferential binding of transcription factors to relaxed dsDNA substrate was observed using DNA origami frame to organize DNA binding domains. Reproduced with permission from ref. [46]. Copyright 2014, American Chemical Society. (C) Binding interactions between nucleosome proteins was studied on a DNA origami hinge device, in which changes in the angle between the two arms would be used to deduce nucleosome interactions. Reproduced with permission from ref. [51]. Copyright 2016, the authors.

The first study utilizing DNA origami nanostructures to study molecular interaction was published in 2008, as Rinker et al. reported the use of DNA origami to study the distance-dependency of multivalent protein–ligand interactions (Figure 4.4A) [42]. Multivalency is a characteristic of certain receptor–ligand interactions in which simultaneous binding of multiple ligands to a single receptor is used to strengthen or modulate the binding affinity. DNA origami nanostructures, with their high spatial resolution and global addressability, thus provide an excellent tool for understanding how the distance between ligands alters the multivalent binding properties at the single-molecule level. The well-characterized thrombin-binding aptamer was used as a model ligand, primarily due to the ease of incorporating it into the DNA nanostructure. The authors utilized a rectangular DNA origami nanostructure to organize pairs of aptamers spaced either ~20 nm or ~6 nm apart. Following thrombin addition, the authors were able to visualize approximately a 10-fold increase of protein binding to the pair of aptamers spaced ~6 nm apart relative to those spaced ~20 nm apart, which confirmed the distance-dependency of these multivalent molecular interactions.

The kinetics of protein–ligand interactions, rather than the spatial relationships, have also been studied using DNA origami platforms. Quantitative measurements of the binding process, such as binding rate and dissociation constant, help to characterize receptor–ligand interactions, and are vitally important for drug development, among other things. Reporting in 2011, Wu et al. describe the use of DNA origami as a platform to facilitate single-molecule characterization of the binding rate for a model receptor-ligand system, streptavidin-biotin [43]. The authors utilized the 2D rectangular DNA origami nanostructure and time-lapse AFM imaging in order to visualize streptavidin binding to biotinylated oligonucleotides. The authors were able to show that the binding-process occurs in two stages with different binding rates, and that near-complete binding could be achieved after 30 minutes with relatively low streptavidin concentrations. Continual measurements after washing away excess protein allowed the authors to study the dissociation rate and consistent with previous results, they found no dissociation after 30 minutes of imaging. Another study looked at reversible streptavidin binding in a dynamic system involving desthiobiotin, which has a lower binding affinity for streptavidin than biotin and thus exhibits reversible binding in the presence of excess free biotin [44]. Using a 2D rectangular origami nanostructure modified with biotinylated and desthiobiotinylated-staples, the authors found almost all of the biotin and desthiobiotin sites were occupied by streptavidin but selective loss of binding at only the desthiobiotin sites occurred following addition of excess biotin.

Work by Subramani et al. in 2010 investigated secondary DNA binding sites of human topoisomerase I (hTopoIB) using a DNA origami platform [45]. In this study, the authors used the canonical 2D rectangular DNA origami nanostructure with a single dsDNA fragment extending from the DNA surface to monitor binding of human topoisomerase 1. The addition of both purified hTopoIB and hTopoIB-dsDNA conjugates with DNA origami nanostructures led to appearance of bright spots in AFM images on a portion of the structures, and this was validated as hTopoIB binding to both one and two dsDNA helices.

One major benefit of using DNA origami as a platform for detecting or studying single-molecule interactions is the modularity of the design—once the basic structure is in place, mere changes in a small number of strands allows for a wide variety of chemical or biochemical species to be studied. This advantage is most clearly demonstrated in the work of Hiroshi Sugiyama's group, who developed a single DNA origami chip that has been used to study many different single-molecule processes. In the original paper by Endo et al., the authors designed a 2D rectangular DNA origami structure with a square hole in the center, which they call a "DNA frame" [30]. Four different ssDNA extensions extend into the middle of the hole, providing anchoring points for loading various cargos. The DNA origami frame allows for easy AFM imaging and added stability to the cargo, while the hole in the center facilitates high resolution AFM imaging by limiting the amount of background near the targets. It should be noted that the Sugiyama group has used small variations of this design to study a huge array of single-molecule events—there will be similar iterations of this DNA origami frame throughout the rest of the chapter.

In a report published in 2014, this group used the DNA origami frame construct to study the binding of two transcription factors, Sox2 and Pax6, to regulatory DNA elements (Figure 4.4B) [46]. In particular, the authors sought to understand how DNA topology (i.e., bending) would contribute to protein-complex binding. The authors loaded two different dsDNA helixes of different lengths containing the Sox2 binding sequence into the DNA origami frame: a 64-nt strand which just fits inside the frame and a 74-nt strand which can exhibit some degree of flexibility. The authors found that DNA topology does affect Sox2 binding, as AFM imaging of DNA origami constructs incubated with Sox2 found higher protein occupancy on the relaxed strand than on the tensed strand.

While the majority of single-molecule binding studies focus on protein–ligand interactions, experiments analyzing interactions between DNA molecules can also provide useful information, particularly for biotechnology applications utilizing DNA nanotechnology. In a study published in 2010, Jungmann et al. use single-molecule fluorescence to study the kinetics of DNA hybridization on DNA origami nanostructures [47].

The authors designed a narrow, 2D DNA origami ribbon with several staples extended to present "docking" strands. Using total internal fluorescence microscopy (TIRF), they were able to determine when fluorescently labeled strands complementary to docking sequences bound to the origami structure, which facilitated the calculation of reaction constants kon and koff for strands with different lengths and at different temperatures.

Researchers from Sugiyama's group published similar work in 2012, studying the hybridization between photoresponsive DNA oligonucleotides [48]. These modified oligonucleotides contain artificial nucleotides with azobenzene moieties—irradiation with UV and visible light induces transitions between *trans*- and *cis*-isomers, which facilitates transition between hybridized (trans) and free (cis) DNA strands. The authors used the DNA origami frame design to load two parallel DNA duplexes containing azobenzene-modified ssDNA extensions in their center, such that hybridization/destabilization of the two strands leads to the formation of X-shaped/parallel architectures. The kinetics of switching between these two states following irradiation with UV and visible light was analyzed using HS-AFM imaging, together with the effect of temperature on transition yields.

Alternative nucleic acids have gained considerable interest in recent years, primarily due to their enhanced stability and resistance to nuclease degradation. In a report published in 2012, Yamazaki et al. used DNA origami to investigate peptide nucleic acid (PNA) invasion into DNA duplexes at the single molecule level [49]. PNAs consist of nucleobases linked to each either through amide (peptide) bonds rather than phosphodiester bonds. The neutral-charged PNAs have stronger binding to complementary DNA due to reduced electrostatic repulsion, which leads to PNA invasion into DNA duplexes, in which PNA can displace complementary DNA strands without the need for a toe-hold. In this study, the authors used a DNA origami plier device that had been developed previously for the detection of various chemical and biologic molecules (see Section 4.2) in order to study PNA–DNA invasion. Incubation with complementary PNA molecules led to dissociation of locking strands and transition from a parallel to an X-shaped conformation, while single mismatches in PNA molecules inhibited the transition.

There have been a series of recent studies that have further expanded the use of DNA origami nanostructures for analysis of molecular interactions. These nanostructures have been used to measure and quantify the forces acting to hold these interactions together. In one study, Kilchherr et al. developed a technique to study the stacking forces between adjacent DNA bases [50]. The authors developed a system with two rigid DNA origami rods connected by a long ssDNA tether with a predefined number of cantilevered blunt-end DNA duplexes facing inwards from the two rods. By manipulating optical tweezers to control the distance between the two rods, the blunt end duplexes could be positioned close enough together to enable base-pair stacking and then pulled apart to monitor the force at which this interaction was broken. The modularity of the DNA origami design allowed the authors to investigate how stacking forces were dictated by each specific base–base interaction (i.e., G:T, C:A, etc.), as well as how the number of stacked duplexes affected the force landscape.

Another study by the same group used a different DNA nanostructure to probe the interaction between two nucleosomes (Figure 4.4C) [51]. In this work, Funke et al. used a previously designed force spectrometer [52], which consists of two rigid DNA rods linked at one end by a molecular hinge such that each rod could move either toward or away from each other around this hinge. The authors included ssDNA tethers at the hinge, which act as entropic springs to limit rod movement and enable the extraction of energy landscape information from positional data measured using TEM. By attaching the nucleosomes to both arms at different distances from the hinge and in different orientations, the authors were able to use differences in positional frequency relative to free structures to infer the free energy landscapes of nucleosome interactions.

A very similar study on the nucleosome stability was conducted by Le et al., though they were interested primarily in intranucleosome interactions [53]. The authors used a very similar hinge-based DNA origami nanocaliper, but linked either end of the core histone protein within the nucleosome to each arm of the device. In this way, the length and conformation of the nucleosome could be monitored by measuring the angle at the hinge. By controlling the length of excess linker DNA between nanodevice arms and the nucleosome, the winding state of each nucleosome could be estimated based on expected and calculated end-to-end distances, and the authors were able to monitor nucleosome stability and transcription factor binding through measurement of the nanocaliper angle.

4.3.1 Protein Activity

Proteins are the machines of cells, and play at least some role in virtually every critical biological process. While proteins and the role they play in cellular physiology and pathophysiology have been studied in bulk ensemble measurements since the advent of the molecular biology revolution in the 1950s, the ability to study the dynamic activity of proteins at the single-molecule level promises to further increase our base of knowledge.

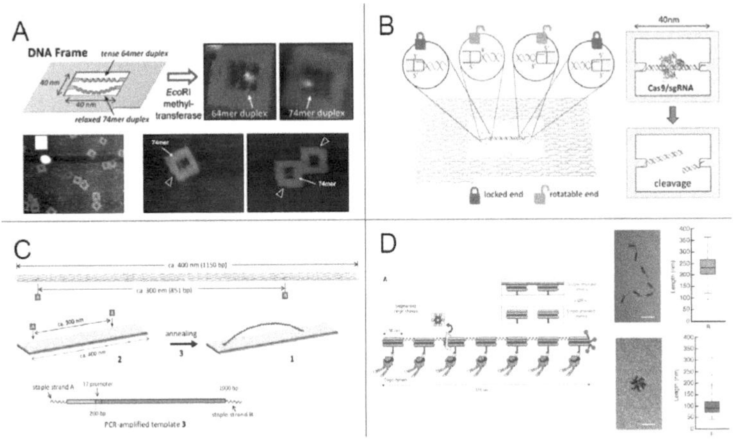

Figure 4.5 Single-molecule analysis of protein activity. (A) Activity of the protein EcoR1 methyl-transferase was preferentially observed on relaxed dsDNA substrate localized with DNA origami frame. Reproduced with permission from ref. [30]. Copyright 2010, American Chemical Society. (B) Identification of preference for Cas9 protein to cleave target DNA based on torsional constraints of DNA duplex using DNA origami frame. Reproduced with permission from ref. [29]. Copyright 2016, American Chemical Society. (C) The new DNA origami nanostructure facilitates the single-molecule studies of transcription with T7 RNA polymerase (RNAP). Reproduced with permission from ref. [58]. Copyright 2012, Wiley-VCH. (D). The new DNA origami design consisted of seven rigid 12HB segments, linked together by stretches of flexible ssDNA used to probe whether cargo rigidity affected motor protein behavior. Reproduced with permission from ref. [62]. Copyright 2016, Wiley-VCH.

The benefits of DNA origami for single-molecule studies discussed previously are especially useful for studying protein activity at the single-molecule level. The well-defined coordinate space of DNA origami nanostructures can be used to track the motion of DNA-binding proteins over time. The ability to precisely control the orientation and topology of various components, such as proteins is infeasible in other biochemical assays, while the ability to distinguish different recombinant products via their anchoring points allows for unique hypotheses to be tested. The spatial addressability of the DNA origami platform allows for the precise incorporation of heterogeneous mixtures of proteins in well-defined ratios and the ability to incorporate many fluorescent dyes per structure leads to higher signal to noise ratios in optical microscopy studies. However, there is one notable limitation—real-time studies using AFM on DNA origami platforms mean protein activity is being visualized on a solid-surface, which can differ greatly from solution-based processes. For example, one study calculated slower diffusion constants than those reported in solution, which is probably due to interactions with the mica surface.

One of the first uses of DNA origami to study enzymatic activity was demonstrated in 2010, in which researchers from the Sugiyama group first used their DNA origami frame [30]. In particular, the authors were interested in studying how structural properties of dsDNA (i.e., tensed versus relaxed) effects the activity of enzymes hypothesized to deform, or bend DNA. The initial study focused on methylation of DNA by the enzyme *Eco*R1 methyltransferase (M.*Eco*R1) (Figure 4.5A). Two different dsDNA substrates were loaded into the frame: a short, tensed sequence which just fits within the DNA origami frame, and a long, relaxed sequence which can accommodate up to a 60° bend at the target sequence. Using HS-AFM imaging, the authors found that M.*Eco*R1 bound preferentially to the relaxed sequence rather than to the short sequence, and dynamic imaging revealed the movement of the enzyme along the substrate. Similar work on the effect of tensed versus relaxed DNA conformation on base excision repair through the activity of two enzymes, 8-oxoguanine glycosylase (hOgg1) and T4 pyrimidine dimer glycosylase, was conducted by the same group [54]. The authors loaded long and short dsDNA targets containing damaged nucleobases into the DNA origami frame. In order to visualize enzyme activity, both duplexes contained a nick in the strand opposing the damaged nucleobases; enzymatic single-strand breaks at the damaged base leads to double-strand breaks visible on AFM images. The authors found that both enzymes preferentially repaired the relaxed duplex, and real-time imaging with HS-AFM enabled the visualization of protein motion along the dsDNA duplex until the enzymes reached the position of damaged base.

In 2014, the Sugiyama group used the DNA origami frame to study site-specific recombination at the single-molecule level, specifically how the orientation of loxP substrates effects the activity of Cre recombinase [55]. In order to study site-specific recombination in real time, a slight change was made in the DNA origami frame design; the anchoring points for loading substrates were positioned such that two perpendicular pairs extended into the center of the frame. Two loxP DNA sequences could be loaded into the frame, such that the ends of the dsDNA were linked to adjacent sides to form loops and duplexes did not overlap with each other. This design allowed the authors to control the orientation of one of the dsDNA segments, in order to compare the differences between loxP arranged in parallel versus antiparallel orientations. AFM imaging of DNA frames following addition of Cre recombinase was used to verify the formation of the synaptic complex (an X-shaped duplex architecture with a bright spot in the center) as well as successful recombination (switch between neighboring anchoring points connected by loops). HS-AFM imaging of recombination events in real time showed that Cre could bind to both antiparallel and parallel loxP orientations but recombination could only occur in the antiparallel conformation. Further changes to substrate orientations enabled studies to investigate preferential cleavage at A-T versus G-C sites as well as effects of preformed intermediate state topology on final resolution.

The Sugiyama group also investigated the activity of another enzyme involved in homologous recombination, RecU resolvase [56]. The authors used the DNA origami frame design which allows for the incorporation of two parallel DNA double-helices, and these anchoring points were used to attach both tensed and relaxed antiparallel four-way junctions into the origami frame. AFM imaging showed equal binding of RecU but preferential resolution in the relaxed substrate. Finally, the specific design used in this report allowed for the distinction between two different possible cleavage outcomes, as one resolution pathway leads to two parallel dsDNA while the second leads to two dsDNA loops.

The Sugiyama group continued to use their DNA frame to study the interactions of proteins with DNA, focusing their attention on the CRISPR-Cas9 system in a publication in 2016, studying how the torsional constraints on DNA would affect Cas9 activity (Figure 4.5B) [29]. They used a DNA origami frame design with a single duplex spanning the center cavity into which four different ssDNA strands could be loaded. This allowed for four different nanoframe designs; (1) Both target and nontarget complementary DNA strands were bound to both edges, thus providing torsional constraint, (2–3) where either the target or nontarget strand was only bound to one edge, providing some flexibility, and (4) where both target and nontarget strands were only bound to one edge. AFM imaging of all samples incubated with Cas9 and guide RNA showed that the kinetics of Cas9 binding were affected by strand topology—it took significantly longer for the protein to bind to frames where the nontarget strand was constrained. Quantitative PCR was used to verify that cleavage was hindered on those frames, while HS-AFM enabled the tracking of the cleavage reaction in real time. This study is an excellent example of how the modularity and precision of DNA nanostructures can be used to ask scientific questions, queries that would be unfeasible with more standard techniques.

Other researchers outside of the Sugiyama group have also sought to use DNA origami nanostructures to study DNA repair at the single-molecule level. In a report by Tintore et al. in 2013, the authors developed a DNA origami nanodevice which was used to detect the activity of the enzyme human O6-alkylguanine-DNA alkyltransferase (hAGT), an enzyme which repairs O6-guanine adducts [57]. In this work, the authors used the canonical 2D rectangular DNA origami template to immobilize G-quadruplex-based aptamers toward α-thrombin, such that G-quadruplexes in the correct conformation (i.e., with repaired O6-guanine adducts) bind to α-thrombin, which produces a detectable AFM signal on the DNA origami frame. The authors incorporated the sensors into two double rows of G-quadruplexes on the surface of the DNA origami nanostructure—a control row of unmodified G-quadruplexes which bind to α-thrombin, and a test row containing methylated-guanine residues which can't form q-quadruplexes. Incubation with α-thrombin led bright spots in AFM images only on unmodified aptamers, while preincubation with hAGT and damaged staple strands prior to DNA nanostructure annealing led to α-thrombin binding to bow rows.

In 2012, the Sugiyama group designed a new DNA origami nanostructure to facilitate the single-molecule studies of transcription with T7 RNAP (Figure 4.5C) [58]. This new "observation platform" consisted of a 6HB rectangular DNA nanostructure ~350 nm in length with anchoring points for cargo at either end, which can be used to study processes on DNA sequences longer than can be incorporated into the original DNA origami frame. The authors attached a 1,000 bp template dsDNA containing the T7 promoter sequence to both anchoring points, such that the template dsDNA spanned the length of the DNA origami structure. Movement of the RNAP along the dsDNA template could be seen using HS-AFM, and the authors were able to calculate a 1D diffusion constant from the images. By using biotinylated nucleotide triphosphate precursors, the authors were able to visualize the nascent RNA transcript by labeling with streptavidin as the single-strand RNA product did not produce substantial AFM contrast.

A report published by Okholm et al. in 2015 demonstrated the use of DNA origami as a platform for the single-molecule visualization of DNA polymerization using the enzyme terminal deoxynucleotidyl transferase (TdT) [59]. The authors used the canonical 2D rectangular DNA origami array in which 3 staples contain single stranded overhangs. Polymerization from these staples could be visualized as the appearance of brighter spots on the nanostructure, thus eliminating the need in the previous study for streptavidin labeling. The authors found that elongation of the staples only occurred when bovine serum albumin was added prior to TdT and nucleotide precursor addition. Chao et al. expanded upon this work, using a DNA origami frame and time-lapse AFM imaging to study the activity of a bioactive fragment of DNA polymerase 1 from Escherichia coli (Klenow fragment) [60]. Using a triangular DNA origami frame with an ssDNA template spanning an empty central cavity, the authors were able to visualize protein binding and subsequent movement. The use of an asymmetric frame enabled analysis of directional polymerase movement, in the expected 5'→ 3' direction. AFM imaging was also used to verify formation of a dsDNA product by monitoring height changes.

In a report published 2012, Derr et al. used a DNA origami scaffold to study the activity of the motor proteins dynein and kinesin-1 [61]. While single-molecule studies using DNA origami primarily use the nanostructure as a scaffold for immobilization of target molecules and interactions, the DNA nanostructure in this study is utilized as the cargo molecule. The authors designed a rod-like nanostructure, which consists of an inner core and an outer corona. Motor proteins covalently modified with ssDNA could be docked to sites on the outer corona through hybridization with staple extensions, while the inner core was used to label the structure with fluorescent dyes. The author's tracked movement of the DNA nanostructure cargo containing different numbers of motor-proteins along microtubules tracks using TIRF and they were able to show that the number of attached motor proteins had a significant effect on cargo velocity and total run length. They also incorporated combinations of both kinesin and dynein to study the interplay between opposite polarity-directed movement, and found that all ensembles moved unidirectionally with slower velocities and higher probability of immobility suggesting a "tug-of-war" between motor proteins, with dynein typically dominating the movement.

A subsequent study using a modified DNA nanostructure was used to probe whether cargo rigidity affected motor protein behavior (Figure 4.5D) [62]. The new DNA origami design consisted of seven rigid 12HB segments, linked together by stretches of flexible ssDNA, and the compliance could be reduced by adding complementary strands to form dsDNA linkers. Using TIRF imaging to track the movement of the dynein ensembles, the authors found the compliant cargo moved with significantly higher velocities. This study demonstrates how the modularity of DNA origami, in this case designed to alter the mechanical properties of the nanostructure, allows for novel scientific questions to be studied.

4.3.2 Alternative Nucleic Acid Conformations

The ability for nucleic acids to adopt alternate configurations to the canonical B-form duplex plays a major role in both biology and DNA nanotechnology. For example, G-quadruplex structures, in which guanine-rich DNA strands can form a square planar structure known as a tetrad due to alternative Hoogsteen hydrogen bonding when stabilized by a monovalent metal cation. These tetrads can then stack to form G-quadruplexes, which find roles in stabilizing telomeres and modulating the expression of various proto-oncogenes in biology, and are thus promising targets for drug treatment of cancer. Furthermore, G-quadruplexes can be used to act as sensing elements for monovalent cations such as K+ in DNA nanotechnology. The widespread prevalence and utility of G-quadruplexes opens up the need for studies to better understand and utilize these structures, and several different single-molecule studies with DNA origami have been undertaken. Other alternative nucleic acid conformations, such as B-form to Z-form DNA transitions and RNA-RNA kissing interactions, have also been studied at the single-molecule level using DNA origami. Many indirect ensemble measurement techniques, such as gel electrophoresis and circular dichroism spectroscopy, have been used to study these structures, but single-molecule experiments enabling their direct visualization have promise to further expand our knowledge. Additionally, the ability to incorporate two heterogeneous sequences into predefined locations opens up the ability to finely control, and thus study, complex systems. The ability to precisely control strand orientation and stoichiometry using DNA origami allow the investigation of scientific questions that are unobtainable using traditional solution-phase ensemble measurements.

Researchers from the Sugiyama group have used their DNA origami frame construct multiple studies on alternative nucleic acid conformation. The first, published in 2010, was used to visualize the formation of G-quadruplex structures (Figure 4.6A) [63]. Two parallel dsDNA duplexes with single-stranded extensions containing G-rich sequences were loaded into the DNA origami frame, such that G-quadruplex formation arising from interactions between G-containing regions on parallel strands lead to the formation of an X-shaped conformation visible using AFM imaging. The authors were able to successfully visualize G-quadruplex formation in real time using HS-AFM in a variety of situations, including several different G-strand sequences and cation identities.

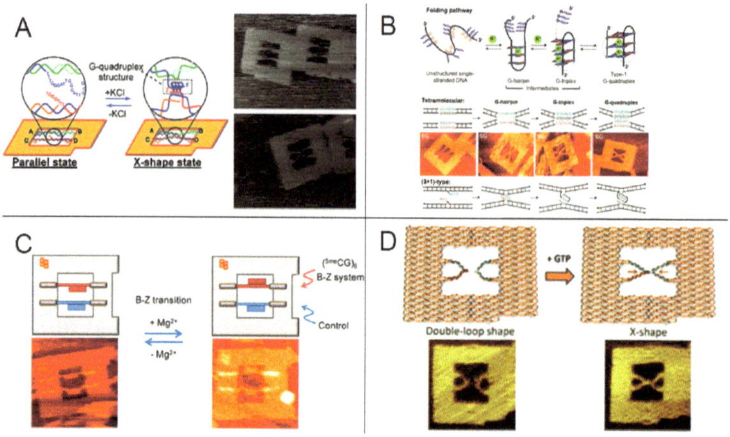

Figure 4.6 Alternative nucleic acid conformations. (A) Formation of G-quadruplex DNA structure was observed within DNA origami frame in presence of potassium ions due to the formation of the X-shape state. Reproduced with permission from ref. [63]. Copyright 2010, American Chemical Society. (B) Formation of G-quadruplex intermediates was observed through careful design of ssDNA extensions inside DNA origami frame. Reproduced with permission from ref. [67]. Copyright 2014, Wiley-VCH. (C) Transition from B-form to Z-form DNA was observed inside DNA origami frame due to transition induced rotation of flag motif on DNA structure. Reproduced with permission from ref. [70]. Copyright 2013, American Chemical Society. (D) Formation of RNA kissing loops in presence of GTP was observed via formation of X-shape structure within DNA origami frame. Reproduced with permission from ref. [72]. Copyright 2015, Royal Society of Chemistry.

In a follow-up study, the Sugiyama group used their DNA origami frame to study the formation of tetramolecular G-quadruplexes, in which G-containing regions from four different strands are used to form the quadruplex [64]. In order to generate a tetramolecular G-quadruplex, the two DNA duplexes loaded into the DNA origami frame each contained a double-stranded region of G–G mismatches, such that a combined four single-stranded G-rich segments were present in an antiparallel configuration. HS-AFM imaging of real-time transitions between X-shape and parallel duplex regions was used to monitor G-quadruplex formation, and the effect of variables such as number of contiguous G-residues, strand polarity, and dsDNA flexibility on kinetics and yield of the assembly could be assessed. Follow-up studies in 2013 and 2014 by the same group investigated the ability for species other than monovalent cations to induce tetramolecular G-quadruplex formation, such as HIV-1 nucleocapsid proteins (NCps) [65] and pyrido-dicarbozamide (PDC) [66]. In both cases, the authors found that NCps and PDC could induce formation of G-quadruplex structures in the absence of potassium ions, and HS-AFM revealed the kinetics of reversible G-quadruplex formation through hypothetical intermediate steps, as well as NCp searching behavior along the duplex.

For their next study, the Sugiyama group sought to better understand the hypothesized G-quadruplex intermediates (Figure 4.6B) [67]. Potential folding pathways to go from unstructured ssDNA to quadruplex structures involving two-stranded G-hairpins and three-stranded G-triplex formation have been implied from previous experiments, but no direct visualization of these intermediates had been previously achieved. Using the DNA origami frame, the authors were able to structurally and stoichiometrically control the DNA strands involved in the G-quadruplex formation, which facilitates more in-depth experiments that can be used to isolate possible intermediate structures. For example, by removing two of the four mismatched G-rich strands in the tetramolecular system, the authors were able to stop G-quadruplex formation at the hypothetical G-hairpin step, while removal of just one of the four mismatch G-rich strands limited assembly to G-triplexes. Using AFM imaging, the authors were able to verify that both G-hairpin and G-triplex structures could form, providing evidence that these structures could exist as intermediates during quadruplex formation.

The Sugiyama group also investigated the conformational changes associated more complex systems involving both G-quadruplex and other nucleic acid conformations. In the first study, the authors also developed a system in which connections between three different duplexes could be modulated. By using both G-quadruplexes and azobenzene-modified oligonucleotides, they were able to study the switching and competition between these two interactions by changing potassium concentrations and the wavelength of irradiating light [68]. In another study, the group investigated the behavior of the promoter sequence insulin-linked polymorphic region (ILPR), which contains a G-rich strand and thus a complementary C-rich strand [69]. While the G-rich strand is capable of forming G-quadruplex structures, the C-rich strand has been shown to form into an i-motif structure. I-motifs are alternative nucleic acid conformations consisting of four DNA strand segments linked together through hemi-protonated cytosine base-pairs. As such, i-motif regions typically form from C-rich strands in acidic environments. The authors included both strands of ILPR as ssDNA regions within two separate duplexes inside the DNA origami frame and were able to study the kinetics of competing processes between duplex formation, G-quadruplex formation in presence of potassium, and i-motif formation at low pH using HS-AFM.

While the Sugiyama group has focused primarily on using their DNA origami frame to study the dynamics of G-quadruplex formation at the single-molecule level, they have also investigated other alternative nucleic acid conformations. In a paper published in 2013, Endo et al. used the DNA origami frame to study the transition between B-form and Z-form DNA (Figure 4.6C) [70]. B-form DNA refers to the classical right-handed helix, while Z-form DNA consists of a left-handed helix with a disjointed, zig-zag pattern of the phosphodiester backbone. Z-form DNA has been shown to occur when bound to certain proteins and along with a biological importance, B-Z transitions lead to mechanical movement of the DNA strands that has been used to construct molecular motors. In this work, the authors loaded DNA constructs into the center of the DNA origami frame which contain a single-duplex "pole" attached to a triple-helix "flag" formed using a small triple-crossover tile. One structure, known as the "rotor," contains a DNA sequence known to exhibit the B-Z transition within the pole, while the second structure is static and used as a control. The flag portion is included to allow for easy identification of the state of the molecular rotor—movement from "flag-down" to "flag-up" conformations identified in AFM images signals transition from B-form to Z-form DNA, and HS-AFM of the glad orientation was used to visualize rotor movement as a function of Mg^{++} concentrations. A subsequent study of B-Z transitions by the Sugiyama group investigated the effect of artificially constrained DNA on Z-form DNA binding proteins [71], in a similar manner as they studied artificially constrained DNA on B-form DNA binding proteins (see Section 3.1). The authors loaded two dsDNA duplexes which just fit inside the DNA origami frame, but with one duplex containing single-strand nicks on regions flanking the CG-repeats which impart additional degrees of freedom. Incubation with the Z-form binding protein Zαβ led to the preferential binding to the strand containing the rotatable CG repeats, which indicates the ability of the artificially constrained DNA sequence to limit B-Z transitions.

In one of their most recent work, the Sugiyama group has used HS-AFM and their DNA origami frame construct to study interactions between alternative nucleic acid conformations known as RNA kissing loops (Figure 4.6D) [72]. Kissing loops are formed from two RNA hairpins which can interact with each other through Watson–Crick base-pairing. This unique RNA structure has been used frequently as a construction element for complex RNA structures, including aptamer-kissing loop sequences which tie ligand-binding behaviors to kissing loop interactions. In particular, the authors focus on a sequence known as the GTPswitch, which can form kissing-loop interactions with its complement component (Aptakiss) only following GTP binding. The authors incorporated the GTPswitch and Aptakiss into the DNA origami frame, such that the terminal ends of each structure were anchored into the same side of the DNA origami frame (i.e., the double-loop architecture). The addition of GTP led to interactions between the two strands and kissing loop formation that was observable by AFM imaging as the X-shaped conformation.

4.4 Control and Visualization of Chemical Reactions

Chemical reactions, narrowly defined here as the process of breaking or forming covalent bonds, have traditionally been studied using ensemble measurements, in which measurements are taken on billions of chemical species which all undergo the transformation. Recent advances in single-molecule techniques allow researchers to move beyond these ensemble measurements to allow for improved analysis of heterogeneous reactions. Furthermore, the detection of chemical reactions, rather than their study, is frequently needed for real-world applications—the same benefits for single-molecule sensing of biomolecules apply here. However, single-molecule analysis of bond formation has traditionally been challenging because it is difficult to know the location of the bond formation prior to the reaction. DNA origami is uniquely situated to studying single-molecule reactions like this, as the local addressability and predefined binding locations enables precise knowledge of reaction sites. Furthermore, the ability to include and identify multiple reactions on the same structure facilitates the analysis of selectivity of each reaction.

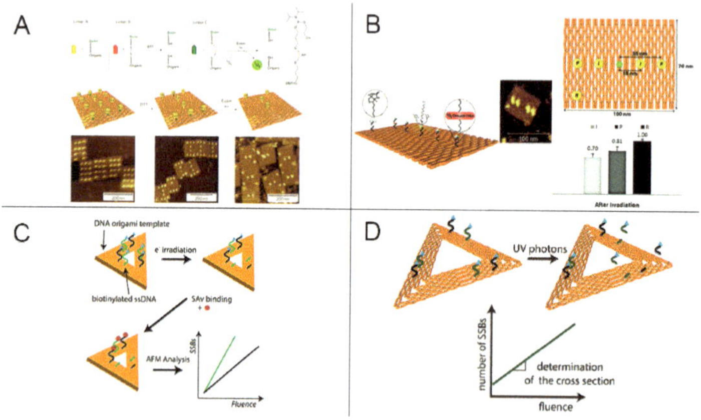

Figure 4.7 Single-molecule analysis of chemical reactions. (A) The formation/cleavage of chemical bonds was tracked on a DNA origami platform by tying chemical reaction to the addition/removal of biotin groups from the surface of the DNA origami structure, which could then be visualized by streptavidin addition using AFM. Reproduced with permission from ref. [73]. Copyright 2010, Nature Publishing Group. (B) Activity of singlet-oxygen monitored via cleavage of singlet-oxygen sensitive bond linking biotin group to DNA origami surface at different positions. Reproduced with permission from ref. [74]. Copyright 2010, American Chemical Society. (C) Electron-induced cleavage of disulfide bonds on ssDNA was monitored via cleavage of biotinylated DNA probed from surface of DNA origami triangular nanostructure. Reproduced with permission from ref. [75]. Copyright 2012, American Chemical Society. (D) UV-induced introduction of single-strand breaks was monitored via cleavage of biotinylated probes on DNA origami surface and was used to calculate the cross-section of UV-DNA interaction. Reproduced with permission from ref. [76]. Copyright 2015, American Chemical Society.

Voigt et al. first demonstrated the use of DNA origami to assist in visualization of single-molecule chemical reactions in 2010 (Figure 4.7A) [73]. The group used the canonical 2D rectangular DNA origami nanostructure as the platform for localizing and viewing the single-molecule reactions using AFM. Streptavidin-biotin binding was used as the sensing element for the biosensor binding of streptavidin to biotin-containing staples producing contrast in AFM images. By engineering biotin incorporation on the staple strands to be dependent on the formation or cleavage of individual chemical bonds, the presence or absence of streptavidin on the DNA origami structure served as a binary reporter on the outcome of the chemical reaction. For bond cleavage reactions, staples were modified with biotin through three chemical groups—a noncleavable amide bond, a reducible disulfide bond, and a singlet-oxygen sensitive 1,2-bis(alkylthio)ethene group. Incubation with dithiothreitol (DTT) led to the removal of all streptavidin at locations containing disulfide linkers, while streptavidin binding to sites containing 1,2-bis(alkylthio)ethene linkers was lost only when a singlet oxygen species was generated using photosensitizers and UV irradiation. The group also studied bond formation. Two different reactions involving chemical groups commonly used in bioconjugation applications were studied: the copper(I)-catalyzed click reaction between alkynes and azides moieties, and reaction between primary amines and N-hydroxysuccinimide (NHS). Staple strands were modified with either azide, amine, or alkyne functional groups and the matching chemical reactants (alkyne, NHS-ester, and azide, respectively), chemically modified with biotin groups were subsequently added. Monitoring streptavidin binding to the various sites allowed the authors to calculate the yield of each reaction.

In a subsequent study, this group also used the same DNA origami structure to study in more detail photosensitized singlet oxygen behavior in a paper published by Helmig et al. (Figure 4.7B) [74]. In this work, the authors incorporated one staple strand conjugated to a single indium pyropheophorbide singlet oxygen photosensitizer (IPS) into the center of the nanostructure and biotinylated staples modified with a singlet oxygen cleavable linker at predefined distances from the IPS. Successful generation of singlet oxygen cleaves the linker, removing the biotin group and preventing streptavidin binding at that location. When modified DNA origami nanostructures were irradiated prior to streptavidin addition, sites lacking streptavidin corresponding to cleaved biotin linkers were found, and the spatial position of the cleavable linker relative to IPS was found to have an effect on the yield of the reaction.

Another group used DNA origami templates to study electron-induced bond cleavage at the single-molecule level in a paper by Keller et al. (Figure 4.7C) [75]. The focus of the study was on low-energy electrons (LEE), which can break and reform chemical bonds. In this work, the authors used a triangular DNA origami template and AFM imaging to quantitatively study the LEE-induced dissociation of disulfide bonds and DNA strand cleavage. As with previous studies, the biotin-streptavidin interaction was used as a read-out for AFM analysis—cleavage of the bond linking biotin to the DNA structure releases biotin, which prevents streptavidin binding to that site. Two groups of biotinylated strands containing either cleavable disulfide linker or stable hydrocarbon linker were used, and the authors were able to calculate the dissociation yield of the electron-induced bond cleavage.

Another study by Vogel et al. in 2015 used DNA origami templates in a single-molecule study to calculate the absolute cross-section for ultraviolet light (UV)-induced strand breakage (Figure 4.7D) [76]. UV-induced DNA damage plays a large role in many genetic human diseases, including cancer. Therefore, the ability to quantify the probability of strand cleavage as a function of DNA sequence and secondary structure is valuable. In this work, the authors used the triangular 2D DNA origami with biotinylated staples for labeling with streptavidin and visualization using AFM. The authors then dried the samples on a substrate and irradiated it with UV light. The authors were able to calculate absolute cross-section of the process by monitoring strand breakage (i.e., loss of streptavidin binding) with increasing photon fluency and found that both photon energy and DNA sequence had an effect.

4.5 Photonic Techniques for Biotechnological Applications

While the majority of single-molecule studies using DNA origami platforms have focused on the analysis of biomolecules and biomolecular interactions, several researchers have focused on studying the behavior of light at the single-molecule level. Photonic techniques, particularly FRET and dye-nanoparticle interactions, notably SERS, and AuNP-induced fluorescence quenching, play important roles in a wide variety of biotechnology applications. As such, the ability to study these processes on the single-molecule level can lead to an improved understanding of the photonic behavior that facilitates the development of these technologies for novel applications.

DNA origami is especially well-suited to studies of photonic behavior at the single-molecule level due to its high spatial resolution, which plays a vital role in these phenomena. The ability to precisely control distances between components in the pathways allow researchers to more intricately explore the mechanisms of photon-matter interactions. Furthermore, the relatively large footprint of DNA origami nanostructures facilitates the immobilization and identification of the processes during the study.

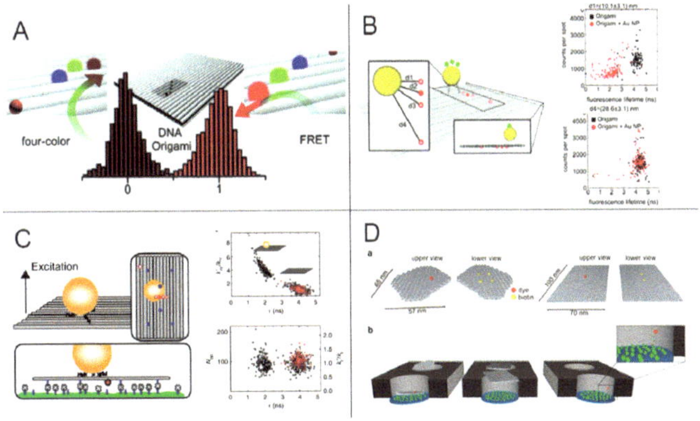

Figure 4.8 Single-molecule analysis of photonic applications. (A) Control of the FRET pathway was achieved by modulating the position of intermediate donor/acceptor dye on the DNA origami structure. Reproduced with permission from ref. [31]. Copyright 2011, American Chemical Society. (B) Distance-dependent quenching of fluorescent dyes from AuNPs monitored on DNA origami structure by controlling the position of a fluorescent dye. Reproduced with permission from ref. [78]. Copyright 2012, American Chemical Society. (C) The DNA origami-AuNP platform to quantify the quantum yield and excitation rate of single fluorescent dyes. Reproduced with permission from ref. [79]. Copyright 2014, Nature Publishing Group. (D) Use of a DNA origami adapter to enable the incorporation of single molecules inside zero mode waveguides. Reproduced from ref. [81]. Copyright 2014, American Chemical Society.

One of the first studies on the use of DNA origami as a platform to study a photonic technique at the single-molecule level was published in 2011 by researchers from the Tinnefeld group (Figure 4.8A) [31]. The authors used the canonical 2D DNA origami rectangle to organize four different fluorescent dyes in precise spatial arrangements in order to realize a four-color FRET pathway. Specific staple strands were covalently modified with four different fluorescent dyes, with a "blue" fluorophore serving as the input dye, a "red" and "IR" fluorophore serving as two possible output dyes, and a "green" fluorophore serving as a jumper dye. The green fluorophore could be placed in two alternate locations, which would facilitate energy transfer to either the red or IR dye based on its spatial arrangement. Measurement of FRET efficiencies using confocal microscopy revealed that energy transfer from input to output dye only occurs when the jumper dye is positioned in between—placing the jumper dye in a different location led to shifts from red dye to IR dye output, while adding jumper dye to both positions at the same time led to both red dye and IR output.

The same group published another study in which they used DNA origami to organize a two-dye FRET system in order to facilitate single-molecule determination of the Forster radius [77]. For this system, the authors used a three-dimensional DNA origami block, designed as a rectangular prism with a height of three helices and a width of 14. This change in design was used because the 3D rectangular prism structure exhibits significantly higher rigidity compared to the 2D rectangle, as studies to determine Forster radii necessitate precise and homogenous interdye distances. The authors were able to calculate the Forster radius of a Cy3/Cy5 donor/acceptor pair by measuring the FRET efficiencies for acceptor dyes placed at nine different distances from donor dye on the DNA origami surface.

In 2012, the Tinnefeld group published a study extending the use of single-molecule DNA origami nanostructures to the analysis of dye-nanoparticle energy transfer pathways, rather than dye–dye transfers [78]. In this work, the authors studied the distance-dependence of a fluorescence quenching energy transfer pathway between a AuNP and a fluorescent nanoparticle using the canonical 2D rectangular origami nanostructure (Figure 4.8B). The authors varied the position a fluorescently-labeled staple strand while keeping the position of the thiolated strands used to capture AuNP constant, such that four different DNA origami nanostructures with different AuNP-dye distances were created. Using a confocal microscopy system in which photons from individual spots (corresponding to single DNA origami nanostructures) are counted, the authors demonstrated that both the intensity and life-time of the fluorescent dye decrease as the distance between the AuNP-dye pair decreases in a manner that closely matches the theoretical models. A continuation of this study was published in 2014, where more advanced microscopy techniques were used in conjunction with the DNA origami-AuNP platform to quantify the quantum yield and excitation rate of single fluorescent dyes [79].

In a report published in *Science* in the same year, the group also reported on the use of DNA origami nanostructures to study the fluorescent enhancement of dye molecules located between two AuNPs at the single-molecule level (Figure 4.8C) [80]. Unlike the quenching phenomenon observed with single metallic nanoparticle-dye systems, the high local electric fields in between two plasmonic nanoparticles can be used to significantly enhance the fluorescence emission of dye molecules in that location. For this purpose, the authors designed a pillar-shaped DNA origami nanostructure consisting of a 220-nm long rod with bulges at one end to provide a stable base for immobilization on a glass slide. In the center of the pillar, two groups of staples were modified to capture 80-nm AuNPs, while a staple strand in the between the two capture sites was modified with a fluorescent dye. The authors found that fluorescent intensities were enhanced when dyes were located between two AuNPs and fluorescent lifetimes were decreased. The design was then adapted to allow for visualization of DNA binding (capture sequences instead of fluorophore-labeled staple at the docking site) as well as Holliday junction fluctuation. This system allowed for higher dye concentrations and increased signal-to-noise ratios, which are needed for single-molecule fluorescence studies, while being substantially simpler and easier to construct than previous nanoantennas built using top-down methods.

In 2014, the Tinnefeld group published a paper exploring the use of DNA origami structures to facilitate the incorporation of single molecules within zero mode waveguides (ZMWs) (Figure 4.8D) [81]. ZMWs are constructed from tiny holes placed in thin (~100 nm) metal films immobilized on glass substrates, and are used to facilitate the single-molecule analysis of biochemical processes, most notably for single-molecule real-time DNA sequencing. ZMWs are currently limited by the ability to load single molecules into each hole as well as the inability to control the position of the reporter molecule within the hole. In this study, the authors constructed a DNA origami nanostructure, termed a "nanoadapter," to facilitate the incorporation of a single-reporter molecule into the center of a ZMW. The nanoadapter was designed as a quasi-elliptical three-dimensional disk constructed from an array of individual DNA origami tubes, which imparts a high degree of structural rigidity compared to 2D nanostructures, and a single dye-labeled staple strand was placed in the center. By using a DNA origami nanostructure as a platform to carry the fluorescent dye, size-exclusion effects were shown to limit adherence of single particles within the ZMW. Furthermore, the authors found that use of the nanoadapters led to a more consistent and homogenous signal between ZMWs, presumably due to the ability of the DNA nanostructures to constrain the location of the dye within the hole.

SERS is another optical technique that is important in numerous biosensing applications. Raman spectroscopy is a technique which monitors vibrational and rotational energy states in molecules using inelastic (or Raman) scattering. The signal generated using this technique is highly dependent on the chemical structure of the analyte, and thus acts as a "fingerprint" to easily identify specific compounds in a label-free manner. Unfortunately, Raman signals are relatively weak, and thus high concentrations of target are needed for bulk measurements. One alternative is to use the electric-field enhancement near the surface of metallic structures, which has been shown to enhance Raman signals by 10^{10} fold and thus enable much higher sensitivity. Recent work has shown that this enhancement factor can be increased even further by the use of electromagnetic "hot-spots" found in the spaces between adjacent metallic nanoparticle dimers [82, 83]. While the nanofabrication of these structures is difficult with traditional top-down methods, DNA origami structures provide an excellent platform for precisely organizing the nanoparticles in 3D space.

While there have been several reports on the use of DNA origami structures to fabricate hot-spots using metallic nanoparticles for SERS applications [84, 85], single-molecule sensitivity has only been achieved in two recent studies. In the first, Prinz et al. used a traditional 2D triangular DNA origami array to scaffold the organization of AuNP dimers in different spatial orientations (Figure 4.9A) [86]. Using fluorescent dye-labeled oligonucleotides as model SERS analytes, the authors achieved single-molecule sensitivity when the dye was placed directly in between the two AuNPs with a silver shell grown in situ. In the second study, Simoncelli et al. used photothermal heating of AuNPs to induce temperature-dependent shrinking of the

DNA origami support, which reduced the distance between the two particles enough to achieve single-molecule sensitivity technique to reduce the distance between nanoparticles without silver growth and still achieve single-molecule sensitivity (Figure 4.9B) [87].

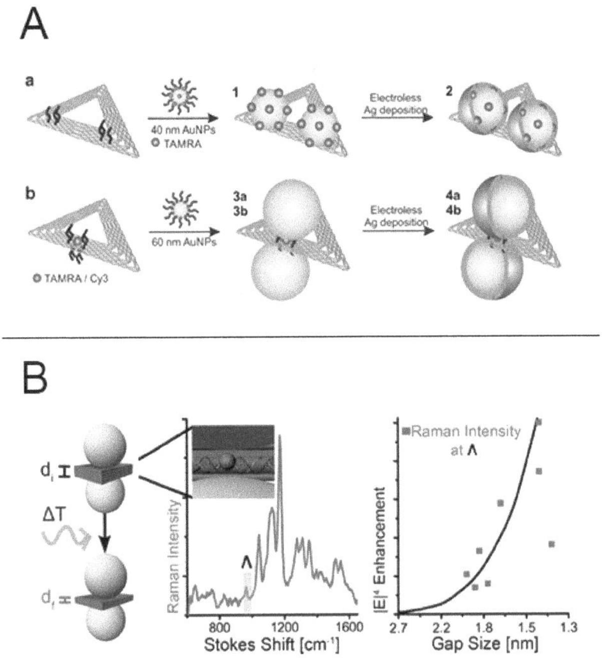

Figure 4.9 Detection of biomolecules using SERS. (A) Enhanced SERS signals for the model analyte were achieved by placing the analyte in between two AuNPs arranged on a DNA origami platform, followed by silver growth on top of gold particles. Reproduced with permission from ref. [86]. Copyright 2016, Royal Society of Chemistry. (B) SERS detection of single-molecule analytes was achieved through photothermal heating of AuNP dimers organized on a DNA origami platform, leading to a shrinking in the gap between the particles. Reproduced with permission from ref. [87]. Copyright 2016, American Chemical Society.

SERS measurements are especially attractive for single-molecule biosensing, as the unique fingerprint for each analyte facilitates the label-free monitoring of a wide variety of target molecules. This strategy has also the benefit of a much higher throughput relative to other techniques such as AFM, TEM, and optical tweezers. However, as is the case with most DNA origami-based biosensing techniques, the current applications are limited due to the need to physically incorporate the target molecule into the DNA nanostructure. While proof of principle studies using fluorescent dyes, which have large Raman scattering cross-sections and are commercially available conjugated to oligonucleotides, are helpful from a mechanistic view point, considerable work is still needed before more practical applications are feasible.

In summary, the ability to control the exact spatial organization of numerous photoactive components, including both fluorophores as well as metallic nanoparticles, makes DNA origami nanostructures an excellent resource for the study of photonic processes like FRET and SERS. Future work is now needed in order to translate the understanding obtained from studies such as these to the development of single-molecule biosensing assays with DNA origami that can be used more cost-effectively in the clinic.

4.6 Summary and Future Perspectives

In the 10 years since Paul Rothemund first introduced DNA origami, the field of structural DNA nanotechnology has rapidly evolved from a neat but niche technology to a promising technique for organizing matter on the nanoscale. Breakthroughs in DNA origami design have led to an ever increasing array of more complex nanostructures, while advances in computational design make it easier than ever for new labs to design DNA nanostructures. However, some of the most exciting advances have come in studies demonstrating the applicability of DNA origami nanostructures to help answer scientific question or solve real-world problems. Applications in single-molecule biosensing in particular have shown considerable promise, as has been highlighted in this chapter. The precision and addressability with which sensing elements or other species can be localized allows for highly detailed studies necessitating accurate spatial organization while also facilitating highly multiplexed bioassays. This same precision also enables the construction of complex structures, giving researchers an impressive tool kit with which to control not only the spatial organization but also the mechanical and physical properties of targets/substrates. DNA origami nanostructures simplify the observation of single-molecule events using techniques such as AFM,

TEM, and optical microscopy by providing stable scaffolding, and the nature of bottom-up self-assembly significantly enhances the fabrication process relative to comparable top-down techniques.

Despite the successes described in this chapter, further advances are needed before the DNA origami platform can realize its full potential for biosensing applications. One of the major limitations is the relative chemical inertness of DNA—as has been seen, the majority of current studies involve nucleic acids or proteins which natively interact with DNA. The ability to incorporate enhanced functionality into these structures is needed to improve the breadth of this technique. There have already been some studies attempting to incorporate other proteins into DNA nanostructures, including the use of SNAP-tags [61] and nickel–NTA interactions [88], but more work is needed to develop simpler, more straightforward techniques. The development of larger DNA origami nanostructures, either through the use of longer scaffold strands or the hierarchical ordering of origami subunits, has the potential to study single-molecule behaviors that occur at longer length scales. The development of more advanced DNA nanostructures which can dynamically control conformation would also open up a new array of possible applications in biosensing. Finally, the practical use of DNA origami nanostructures, especially for traditional sensing applications, is hindered by the experimental techniques used. AFM and TEM, the two most commonly used methods, are expensive, technically challenging to use, and have notoriously low throughput. Improvements in the use optical techniques, particularly moving from proof-of-principle studies on fluorescent dyes to relevant biomolecules, have the potential to significantly advance the use of DNA origami for biosensing.

Appendix: A Tutorial Protocol: Design and Fabrication of 2D Rectangular DNA Origami

Presented here is a brief summary of a generic protocol for designing, synthesizing, and characterizing a simple 2D DNA origami structure. Much more detailed protocols and descriptions can be found in a number of excellent journal articles and other textbooks [7, 89, 90].

Materials

1. **Equipment**
   caDNAno program
   Atomic Force Microscopy
   Agarose Gel Electrophoresis
   Thermocycler
   UV transilluminator
   UV–Vis spectrophotometry
2. **Consumables**
   Pipette tips
   Freeze'N'Squeeze column (Bio-Rad)
3. **Reagents**
   DNA scaffold
   Staple strands
   Tris
   EDTA
   MgCl2
   Ultrapure water (ddH2O)
   Boric acid
   EtBr
   Agarose gel

Methods

- **Step 1: Design of DNA Origami**
  1.1 caDNAno is a program used to assist in the design of DNA origami structures. The software is available for free at caDNAno.org.
  1.2 Open caDNAno and choose square lattice design.
  1.3 In the left panel, click and drag on the open circles in a row to create the number of helices that are desired. Click and drag on the yellow circles a second time to create a starting scaffold strand routing (Figure 4.10A). The approximate width of the 2D rectangle is roughly 2.5 nm/helix.
  1.4 To create a continuous scaffold strand, produce scaffold crossovers by clicking on a scaffold strand and then clicking on the numbered bracket that appears in between two adjacent helices. Shift-click on adjacent break points

(square and arrow end-points) to create a continuous strand. Finally, designate a starting point by inserting a scaffold breakpoint using the break tool (Figure 4.10B).

1.5 Select the scaffold crossovers at the end of the design and drag to extend the helices to the desired length. The length of the 2D rectangle is approximately 0.34 nm/bp of helices.

1.6 Click auto-staple to produce a starting set of staple crossovers, and then click auto-break to enter your desired parameters for staple strands and produce individual strands. The strands can then be manipulated to produce the desired staple domain length and number (Figure 4.10C).

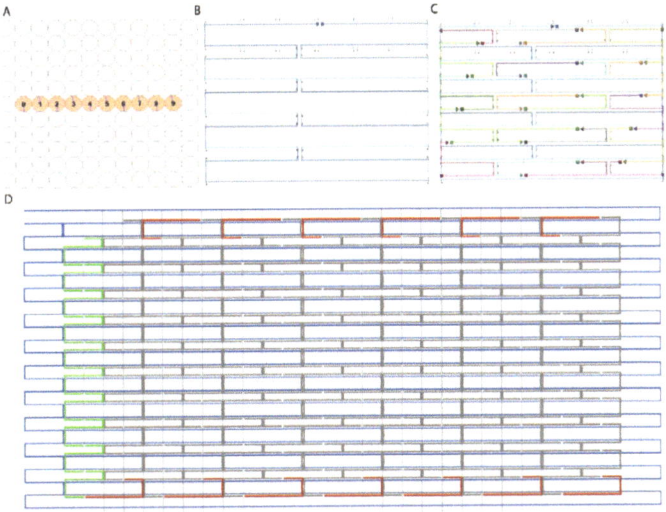

Figure 4.10 caDNAno design of 2D DNA origami rectangle. (A) Screenshot showing set-up for nine parallel helices forming 2D origami rectangle. (B) Screenshot of scaffold strand routing. (C) Screenshot of structure design after auto-staple and auto-break commands, showing design of staple strands. (D) Complete caDNAno design of a full 2D DNA origami rectangle. Scaffold strands are shown in blue, core staple strands are shown in gray, and staple strands to be modified for later applications are shown in green and red. Free scaffold loops on the ends are used to help prevent tile dimerization via base-stacking interactions.

1.7 To assign sequences to your design, click the sequence tool on the right, choose the starting point of the scaffold strand, and then select the desired scaffold sequence. Sequences of individual staples can then be exported to a. csv file using the export tool.

1.8 caDNAno files can be uploaded to the CanDo servers (cando-dna-origami.org) in order to assist in the prediction of the solution-phase structure of the DNA origami design.

1.9 To incorporate modifications such as biotin groups, identify staple strands in the location where the modification should go. In the .csv file, append an additional sequence to the 3' end of the staple, and then purchase a complementary strand with a biotin modification at the 5' end. For example, extend a staple by the sequence 5'Staple-AAAAAAAAAA-3' and then purchase a 5'Biotin-TTTTTTTTTT.

- **Step 2: Synthesis and Purification of DNA Origami**

2.1 Once the staple strands have been designed, oligonucleotides can be ordered from a vendor such as Integrated DNA Technologies. Low quantities (10–25 nmol) and minimal purification (standard desalting) are typically sufficient for most origami structures and applications. Scaffold strands can also be purchased from vendors such as IDT.

2.2 10 nM scaffold strand and 100 nM staple strands (10-fold excess) should be mixed together with annealing buffer (5 mM Tris, 1 mM EDTA, 10 mM $MgCl_2$) in a total volume of 50 μL. Different designs require different magnesium concentrations, and thus it is recommended that a screen of multiple magnesium concentrations between 6 mM and 20 mM be used for the initial assembly.

2.3 Staple/Scaffold mixture should be placed inside a thermocycler and subjected to a thermal annealing protocol. While the length of time is different for each structure depending on complexity, a general annealing protocol for a 3D origami structure is presented: 85°C for 10 minutes, 65°C to 25°C with 1°C decrease every 20 minutes.

2.4 After annealing, the crude scaffold/staple mixture is loaded into a 1.5% agarose gel (gel prepared in 0.5 × TBE buffer - 45 mM Tris, 45 mM Boric acid, 1 mM EDTA, 10 mM $MgCl_2$ and 0.005% (v/v) EtBr) and run at 50 V to 60 V for 1 to 2 hours.

2.5 The band containing the primary DNA origami product can be extracted by cutting out the band using a UV transilluminator. The gel fragment is then crushed and placed into a Freeze'N'Squeeze column (Bio-Rad) and centrifuged to collect the purified DNA origami structure. Concentration of purified DNA origami solution is calculated using UV–Vis spectrophotometry.

2.6 Additional purification methods, such as ultrafiltration and PEG precipitation, can be used when larger quantities of DNA origami samples are required. Details on these purification methods can be found in other references [86, 87].

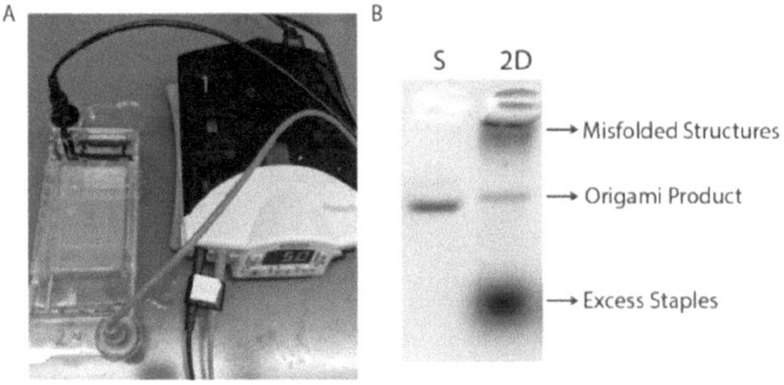

Figure 4.11 Agarose gel electrophoresis (AGE) of DNA origami. (A) Set-up of an agarose gel electrophoresis experiment with 1: Power Supply, 2: Buffer chamber, 3: 1.5% agarose gel. (B) Sa 15mple AGE results for 2D rectangular DNA origami structure comparing blank scaffold strand (S) to folded 2D structure (2D). Bands in origami lane include misfolded/aggregated structures (top), the desired origami product (middle), and excess staple strands (bottom).

- **Step 3: Characterization of DNA Origami**

  3.1 Agarose gel electrophoresis

  3.2 Load ~0.05 pmol of DNA origami into 1.5% agarose gel (prepared in 0.5 × TBE buffer - 45 mM Tris, 45 mM Boric acid, 1 mM EDTA, 10 mM $MgCl_2$, and 0.005% (v/v) EtBr) and run at 50V to 60V for 1 to 2 hours (Figure 4.11A). Image gel under UV Illumination using a gel imager. Correctly formed products should migrate as a single, clean band. Band smearing or accumulation in the well indicates damaged or misformed structures (Figure 4.11B).

  3.3 Atomic force microscopy (AFM)

  3.4 AFM of DNA origami structures is best conducted in solution using a fluid cell but dry in-air imaging is also possible. Load AFM tip (e.g., SNL-10, Bruker) into the fluid cell.

  3.5 Clean mica surface on imaging puck using scotch tape until a single layer of mica remains on the surface.

  3.6 Deposit 10 µL to 20 µL of DNA origami solution onto the mica surface and leave for 1 to 2 minutes for DNA origami to adhere to the surface. Add 50 µL of imaging buffer (5 mM Tris, 1 mM EDTA) and place it into the AFM microscope.

3.7 Follow protocol for individual AFM microscope to collect image. Tapping mode is preferred to minimize damage to DNA origami structure. Figure 4.12 shows an AFM image of 2D rectangular DNA origami.

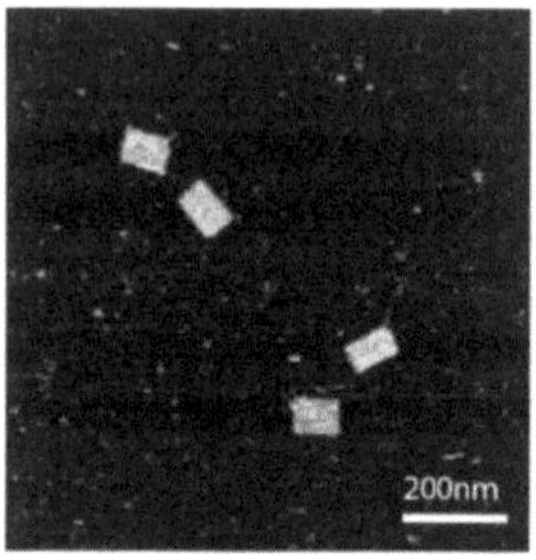

Figure 4.12 Atomic force microscopy of 2D rectangular DNA origami.

References

1. Seeman NC (2003). DNA in a material world, *Nature, 421* (6921), 427–431.
2. Seeman NC (1982). Nucleic acid junctions and lattices, *J Theor Biol*, *99* (2), 237–247.
3. Chen J, Seeman NC (1991). Synthesis from DNA of a molecule with the connectivity of a cube, *Nature*, *350* (6319), 631–633.
4. Ke Y, Ong LL, Shih WM, Yin P (2012). Three-dimensional structures self-assembled from DNA bricks, *Science*, *338* (6111), 1177–1183.
5. Rothemund PWK (2006). Folding DNA to create nanoscale shapes and patterns, *Nature*, *440* (7082), 297–302.
6. Hong F, Zhang F, Liu Y, Yan H (2017). DNA origami: scaffolds for creating higher order structures, *Chem Rev*, *117* (20), 12584–12640.
7. Castro CE, Kilchherr F, Kim DN, et al. (2011). A primer to scaffolded DNA origami, *Nat Methods*, *8* (3), 221–229.
8. Ke Y, Sharma J, Liu M, Jahn K, Liu Y, Yan H (2009). Scaffolded DNA origami of a DNA tetrahedron molecular container, *Nano Lett*, *9* (6), 2445–2447.
9. Andersen ES, Dong M, Nielsen MM, et al. (2009). Self-assembly of a nanoscale DNA box with a controllable lid, *Nature*, *459* (7243), 73–76.
10. Douglas SM, Dietz H, Liedl T, Högberg B, Graf F, Shih WM (2009). Self-assembly of DNA into nanoscale three-dimensional shapes, *Nature*, *459* (7245), 414–418.

11. Ke Y, Douglas SM, Liu M, et al. (2009). Multilayer DNA origami packed on a square lattice, *J Am Chem Soc*, *131* (43), 15903–15908.

12. Dietz H, Douglas SM, Shih WM (2009). Folding DNA into twisted and curved nanoscale shapes, *Science*, *325* (5941), 725–730.

13. Zhang F, Jiang S, Wu S, et al. (2015). Complex wireframe DNA origami nanostructures with multi-arm junction vertices, *Nat Nanotechnol*, *10* (9), 779–784.

14. Veneziano R, Ratanalert S, Zhang K, et al. (2016). Designer nanoscale DNA assemblies programmed from the top down, *Science*, *352* (6293), 1534.

15. Benson E, Mohammed A, Gardell J, et al. (2015), DNA rendering of polyhedral meshes at the nanoscale. *Nature*, *523* (7561), 441–444.

16. Wang P, Gaitanaros S, Lee S, Bathe M, Shih WM, Ke Y (2016). Programming self-assembly of DNA origami honeycomb two-dimensional lattices and plasmonic metamaterials, *J Am Chem Soc*, *138* (24), 7733–7740.

17. Zhao Z, Liu Y, Yan H (2011)., Organizing DNA origami tiles into larger structures using preformed scaffold frames, *Nano Lett*, *11* (7), 2997–3002.

18. Iinuma R, Ke Y, Jungmann R, Schlichthaerle T, Woehrstein JB, Yin P (2014). Polyhedra self-assembled from DNA tripods and characterized with 3D DNA-PAINT, *Science*, *344* (6179), 65–69.

19. Gerling T, Wagenbauer KF, Neuner AM, Dietz H (2015). Dynamic DNA devices and assemblies formed by shape-complementary, non-base pairing 3D components, *Science*, *347* (6229), 1446–1452.

20. Douglas SM, Marblestone AH, Teerapittayanon S, Vazquez A, Church GM, Shih WM (2009). Rapid prototyping of 3D DNA-origami shapes with caDNAno, *Nucleic Acids Res*, *37* (15), 5001–5006.

21. Wang Z-G, Ding B (**2014**). Engineering DNA self-assemblies as templates for functional nanostructures, *Acc Chem Res*, *47* (6), 1654–1662.

22. Linko V, Ora A, Kostiainen MA (2015). DNA nanostructures as smart drug-delivery vehicles and molecular devices, *Trends Biotechnol*, *33* (10), 586–594.

23. Tørring T, Voigt NV, Nangreave J, Yan H, Gothelf KV (2011). DNA origami: a quantum leap for self-assembly of complex structures, *Chem Soc Rev*, *40* (12), 5636–5646.

24. Yang YR, Liu Y, Yan H (2015), DNA nanostructures as programmable biomolecular scaffolds, *Bioconjug Chem*, *26* (8), 1381–1395.

25. Wang P, Meyer TA, Pan V, Dutta PK, Ke Y (2017). The beauty and utility of DNA origami, *Chem*, *2* (3), 359–382.
26. Ke Y, Lindsay S, Chang Y, Liu Y, Yan H (2008). Self-assembled water-soluble nucleic acid probe tiles for label-free RNA hybridization assays, *Science*, *319* (5860), 180–183.
27. Subramanian HKK, Chakraborty B, Sha R, Seeman NC (2011). The label-free unambiguous detection and symbolic display of single nucleotide polymorphisms on DNA origami, *Nano Lett*, *11* (2), 910–913.
28. Li X, Li Y, Hong L (2016). A novel self-assembling DNA nano chip for rapid detection of human papillomavirus genes, *PLoS One*, *11* (10), e0162975.
29. Räz MH, Hidaka K, Sturla SJ, Sugiyama H, Endo M (2016). Torsional constraints of DNA substrates impact Cas9 cleavage, *J Am Chem Soc*, *138* (42), 13842–13845.
30. Endo M, Katsuda Y, Hidaka K, Sugiyama H (2010). Regulation of DNA methylation using different tensions of double strands constructed in a defined DNA nanostructure, *J Am Chem Soc*, *132* (5), 1592–1597.
31. Stein IH, Steinhauer C, Tinnefeld P (2011). Single-molecule four-color FRET visualizes energy-transfer paths on DNA origami, *J Am Chem Soc*, *133* (12), 4193–4195.
32. Zhang Z, Wang Y, Fan C, et al. (2010). Asymmetric DNA origami for spatially addressable and index-free solution-phase DNA chips, *Adv Mater Weinheim*, *22* (24), 2672–2675.
33. Godonoga M, Lin T-Y, Oshima A, et al. (2016). A DNA aptamer recognising a malaria protein biomarker can function as part of a DNA origami assembly, *Sci Rep*, *6*, 21266.
34. Zhu J, Feng X, Lou J, et al. (2013). Accurate quantification of microRNA via single strand displacement reaction on DNA origami motif, *PLoS One*, *8* (8), e69856.
35. Lu Z, Wang Y, Xu D, Pang L (2017). Aptamer-tagged DNA origami for spatially addressable detection of aflatoxin B1, *Chem Commun (Camb)*, *53* (5), 941–944.
36. Zhang Z, Zeng D, Ma H, et al. (2010). A DNA-origami chip platform for label-free SNP genotyping using toehold-mediated strand displacement, *Small*, *6* (17), 1854–1858.
37. Kuzuya A, Sakai Y, Yamazaki T, Xu Y, Komiyama M (2011). Nanomechanical DNA origami 'single-molecule beacons' directly imaged by atomic force microscopy, *Nat Commun*, *2*, 449.
38. Kuzuya A, Sakai Y, Yamazaki T, et al. (2017). Allosteric control of nanomechanical DNA origami pinching devices for enhanced target binding, *Chem Commun (Camb)*, *53* (59), 8276–8279.

39. Ke Y, Meyer T, Shih WM, Bellot G (2016). Regulation at a distance of biomolecular interactions using a DNA origami nanoactuator, *Nat Commun*, *7*, 10935.

40. Nickels PC, Høiberg HC, Simmel SS, Holzmeister P, Tinnefeld P, Liedl T (2016). DNA origami seesaws as comparative binding assay, *Chembiochem*, *17* (12), 1093–1096.

41. Koirala D, Shrestha P, Emura T, et al. (2014). Single-molecule mechanochemical sensing using DNA origami nanostructures, *Angew Chem Int Ed Engl*, *126* (31), 8275–8279.

42. Rinker S, Ke Y, Liu Y, Chhabra R, Yan H (2008). Self-assembled DNA nanostructures for distance-dependent multivalent ligand-protein binding, *Nat Nanotechnol*, *3* (7), 418–422.

43. Wu N, Zhou X, Czajkowsky DM, et al. (2011). In situ monitoring of single molecule binding reactions with time-lapse atomic force microscopy on functionalized DNA origami, *Nanoscale*, *3* (6), 2481–2484.

44. Wong NY, Xing H, Tan LH, Lu Y (2013). Nano-encrypted Morse code: a versatile approach to programmable and reversible nanoscale assembly and disassembly, *J Am Chem Soc*, *135* (8), 2931–2934.

45. Subramani R, Juul S, Rotaru A, et al. (2010). A novel secondary DNA binding site in human topoisomerase I unravelled by using a 2D DNA origami platform, *ACS Nano*, *4* (10), 5969–5977.

46. Yamamoto S, De D, Hidaka K, Kim KK, Endo M, Sugiyama H (2014). Single molecule visualization and characterization of Sox2-Pax6 complex formation on a regulatory DNA element using a DNA origami frame, *Nano Lett*, *14* (5), 2286–2292.

47. Jungmann R, Steinhauer C, Scheible M, Kuzyk A, Tinnefeld P, Simmel FC (2010). Single-molecule kinetics and super-resolution microscopy by fluorescence imaging of transient binding on DNA origami, *Nano Lett*, *10* (11), 4756–4761.

48. Endo M, Yang Y, Suzuki Y, Hidaka K, Sugiyama H (2012). Single-molecule visualization of the hybridization and dissociation of photoresponsive oligonucleotides and their reversible switching behavior in a DNA nanostructure, *Angew Chem Int Ed Engl*, *51* (42), 10518–10522.

49. Yamazaki T, Aiba Y, Yasuda K, et al. (2012). Clear-cut observation of PNA invasion using nanomechanical DNA origami devices, *Chem Commun (Camb)*, *48* (92), 11361–11363.

50. Kilchherr F, Wachauf C, Pelz B, Rief M, Zacharias M, Dietz H (2016). Single-molecule dissection of stacking forces in DNA, *Science*, *353* (6304).

51. Funke JJ, Ketterer P, Lieleg C, Schunter S, Korber P, Dietz H (2016). Uncovering the forces between nucleosomes using DNA origami, *Sci Adv*, *2* (11), e1600974.

52. Funke JJ, Dietz H (2016). Placing molecules with Bohr radius resolution using DNA origami, *Nat Nanotechnol*, *11* (1), 47–52.

53. Le JV, Luo Y, Darcy MA, et al. (2016). Probing nucleosome stability with a DNA origami nanocaliper, *ACS Nano*, *10* (7), 7073–7084.

54. Endo M, Katsuda Y, Hidaka K, Sugiyama H (2010). A versatile DNA nanochip for direct analysis of DNA base-excision repair, *Angew Chem Int Ed Engl*, *49* (49), 9412–9416.

55. Suzuki Y, Endo M, Katsuda Y, Ou K, Hidaka K, Sugiyama H (2014). DNA origami based visualization system for studying site-specific recombination events, *J Am Chem Soc*, *136* (1), 211–218.

56. Suzuki Y, Endo M, Cañas C, et al. (2014). Direct analysis of Holliday junction resolving enzyme in a DNA origami nanostructure, *Nucleic Acids Res*, *42* (11), 7421–7428.

57. Tintoré M, Gállego I, Manning B, Eritja R, Fàbrega C (2013). DNA origami as a DNA repair nanosensor at the single-molecule level, *Angew Chem Int Ed Engl*, *52* (30), 7747–7750.

58. Endo M, Tatsumi K, Terushima K, et al. (2012). Direct visualization of the movement of a single T7 RNA polymerase and transcription on a DNA nanostructure, *Angew Chem Int Ed Engl*, *51* (35), 8778–8782.

59. Okholm AH, Aslan H, Besenbacher F, Dong M, Kjems J (2015). Monitoring patterned enzymatic polymerization on DNA origami at single-molecule level, *Nanoscale*, *7* (25), 10970–10973.

60. Chao J, Zhang P, Wang Q, et al. (2016). Single-molecule imaging of DNA polymerase I (Klenow fragment) activity by atomic force microscopy, *Nanoscale*, *8* (11), 5842–5846.

61. Derr ND, Goodman BS, Jungmann R, Leschziner AE, Shih WM, Reck-Peterson SL (2012). Tug-of-war in motor protein ensembles revealed with a programmable DNA origami scaffold, *Science*, *338* (6107), 662–665.

62. Driller-Colangelo AR, Chau KWL, Morgan JM, Derr ND (2016). Cargo rigidity affects the sensitivity of dynein ensembles to individual motor pausing, *Cytoskeleton (Hoboken)*, *73* (12), 693–702.

63. Sannohe Y, Endo M, Katsuda Y, Hidaka K, Sugiyama H (2010). Visualization of dynamic conformational switching of the G-quadruplex in a DNA nanostructure, *J Am Chem Soc*, *132* (46), 16311–16313.

64. Rajendran A, Endo M, Hidaka K, Tran PLT, Mergny J-L, Sugiyama H (2013). Controlling the stoichiometry and strand polarity of a tetramolecular G-quadruplex structure by using a DNA origami frame, *Nucleic Acids Res*, *41* (18), 8738–8747.

65. Rajendran A, Endo M, Hidaka K, et al. (2013). HIV-1 nucleocapsid proteins as molecular chaperones for tetramolecular antiparallel G-quadruplex formation, *J Am Chem Soc*, *135* (49), 18575–18585.

66. Rajendran A, Endo M, Hidaka K, et al. (2014). G-quadruplex-binding ligand-induced DNA synapsis inside a DNA origami frame, *RSC Adv*, *4* (12), 6346.

67. Rajendran A, Endo M, Hidaka K, Sugiyama H (2014). Direct and single-molecule visualization of the solution-state structures of G-hairpin and G-triplex intermediates, *Angew Chem Int Ed Engl*, *53* (16), 4107–4112.

68. Yang Y, Endo M, Suzuki Y, Hidaka K, Sugiyama H (2014). Direct observation of the dual-switching behaviors corresponding to the state transition in a DNA nanoframe, *Chem Commun (Camb)*, *50* (32), 4211–4213.

69. Endo M, Xing X, Zhou X, et al. (2015). Single-molecule manipulation of the duplex formation and dissociation at the G-quadruplex/i-motif site in the DNA nanostructure, *ACS Nano*, *9* (10), 9922–9929.

70. Rajendran A, Endo M, Hidaka K, Sugiyama H (2013). Direct and real-time observation of rotary movement of a DNA nanomechanical device, *J Am Chem Soc*, *135* (3), 1117–1123.

71. Endo M, Inoue M, Suzuki Y, et al. (2013). Regulation of B-Z conformational transition and complex formation with a Z-form binding protein by introduction of constraint to double-stranded DNA by using a DNA nanoscaffold, *Chem Eur J*, *19* (50), 16887–16890.

72. Takeuchi Y, Endo M, Suzuki Y, et al. (2015). Single-molecule observations of RNA-RNA kissing interactions in a DNA nanostructure, *Biomater Sci*, *4* (1), 130–135.

73. Voigt NV, Tørring T, Rotaru A, et al. (2010). Single-molecule chemical reactions on DNA origami, *Nat Nanotechnol*, *5* (3), 200–203.

74. Helmig S, Rotaru A, Arian D, et al. (2010). Single molecule atomic force microscopy studies of photosensitized singlet oxygen behavior on a DNA origami template, *ACS Nano*, *4* (12), 7475–7480.

75. Keller A, Bald I, Rotaru A, Cauët E, Gothelf KV, Besenbacher F (2012). Probing electron-induced bond cleavage at the single-molecule level using DNA origami templates, *ACS Nano*, *6* (5), 4392–4399.

76. Vogel S, Rackwitz J, Schürman R, et al. (2015). Using DNA origami nanostructures to determine absolute cross sections for UV photon-induced DNA strand breakage, *J Phys Chem Lett*, *6* (22), 4589–4593.

77. Stein IH, Schüller V, Böhm P, Tinnefeld P, Liedl T (2011). Single-molecule FRET ruler based on rigid DNA origami blocks, *ChemPhysChem*, *12* (3), 689–695.
78. Acuna GP, Bucher M, Stein IH, et al. (2012). Distance dependence of single-fluorophore quenching by gold nanoparticles studied on DNA origami, *ACS Nano*, *6* (4), 3189–3195.
79. Holzmeister P, Pibiri E, Schmied JJ, Sen T, Acuna GP, Tinnefeld P (2014). Quantum yield and excitation rate of single molecules close to metallic nanostructures, *Nat Commun*, *5*, 5356.
80. Acuna GP, Möller FM, Holzmeister P, Beater S, Lalkens B, Tinnefeld P (2012). Fluorescence enhancement at docking sites of DNA-directed self-assembled nanoantennas, *Science*, *338* (6106), 506–510.
81. Pibiri E, Holzmeister P, Lalkens B, Acuna GP, Tinnefeld P (2014). Single-molecule positioning in zeromode waveguides by DNA origami nanoadapters, *Nano Lett*, *14* (6), 3499–3503.
82. Qin L, Zou S, Xue C, Atkinson A, Schatz GC, Mirkin CA (2006). Designing, fabricating, and imaging Raman hot spots, *Proc Natl Acad Sci USA*, *103* (36), 13300–13303.
83. Camden JP, Dieringer JA, Wang Y, et al. (2008). Probing the structure of single-molecule surface-enhanced Raman scattering hot spots, *J Am Chem Soc*, *130* (38), 12616–12617.
84. Prinz J, Schreiber B, Olejko L, et al. (2013). DNA origami substrates for highly sensitive surface-enhanced raman scattering, *J Phys Chem Lett*, *4* (23), 4140–4145.
85. Thacker VV, Herrmann LO, Sigle DO, et al. (2014). DNA origami based assembly of gold nanoparticle dimers for surface-enhanced Raman scattering, *Nat Commun*, *5*, 3448.
86. Prinz J, Heck C, Ellerik L, Merk V, Bald I (2016). DNA origami based Au-Ag-core-shell nanoparticle dimers with single-molecule SERS sensitivity, *Nanoscale*, *8* (10), 5612–5620.
87. Simoncelli S, Roller E-M, Urban P, et al. (2016). Quantitative single molecule surface-enhanced raman scattering by optothermal tuning of DNA origami-assembled plasmonic nanoantennas, *ACS Nano*, *10* (11), 9809–9815.
88. Shen W, Zhong H, Neff D, Norton ML (2009). NTA directed protein nanopatterning on DNA origami nanoconstructs, *J Am Chem Soc*, *131* (19), 6660–6661.
89. Wagenbauer KF, Engelhardt FAS, Stahl E, et al. (2017). How we make DNA origami, *Chembiochem*, *18* (19), 1873–1885.
90. Ke Y, Wang P (2016). *3D DNA Nanostructure: Methods and Protocols*, Humana Press, New York, p xi, p. 282.

Chapter 5

Single-Molecule Manipulation by Magnetic Tweezers

Zilong Guo and Hu Chen

Research Institute for Biomimetics and Soft Matter,
Fujian Provincial Key Lab for Soft Functional Materials Research,
Department of Physics, Xiamen University, Xiamen 361005, China
chenhu@xmu.edu.cn

Single-molecule manipulation techniques apply mechanical constraints to biomolecules, such as proteins or nucleic acids, to study their conformation transitions and interactions. Among all single-molecule manipulation techniques, magnetic tweezers, which possess the capability to apply an intrinsic constant force, are superior in studies when a large dynamic range of force magnitude and time scale is needed. In this chapter, we introduce the basic principles of magnetic tweezers and technical details to build the setup. Researches on DNA mechanics, DNA processing motors, protein folding/unfolding transitions, and interactions between biomolecules are also summarized, and the emerging development of the technique is demonstrated. Finally, a tutorial protocol to study protein unfolding dynamics with magnetic tweezers is presented

Single-Molecule Tools for Bioanalysis
Edited by Shuo Huang

ISBN 978-981-4800-44-0 (Hardcover), 978-1-003-18913-8 (eBook)
www.jennystanford.com

in detail. We hope the readers will be able to build a magnetic tweezers setup and perform their own measurements with this protocol.

5.1 Introduction

In vivo, biomolecules bind to other molecules, cell membranes or subcellular organelles to perform their functions, in which their mechanical properties play important roles [1, 2]. In the process of gene expression, transcription factors bind to the regulation sites on a strand of DNA, which results in bending or twisting of the DNA template [3]. In more complicated circumstances, some crosslinker proteins could join two or more molecules together to form higher ordered structures, such as cytoskeletons, focal adhesions, or cell–cell junctions [4–6]. The cytoskeleton network supports the morphological shape of the cell and the tension within the network regulates cell migration, proliferation, and differentiation via force-dependent interactions between the cell's components [7–9]. Though force is crucial in the regulation of cellular activities, it is challenging to directly study the mechanical properties of such biomolecules in the absence of a suitable technique to perform single molecule manipulations [10].

Over the last 50 years, single-molecule manipulation techniques, which aim to measure dynamic signals produced by a single biomolecule, were invented and have been continuously developed [11–14]. These techniques are widely applied in studies of different biophysical systems, such as protein folding and unfolding [15], conformational transitions of DNA or RNA [16, 17], interactions between proteins and DNA/RNA [18], and mechanism of molecular motors [19]. During single-molecule manipulations, mechanical constraints, such as a force, a torque, or their combinations, are applied to a single molecule and the resulting conformational transitions are monitored with a nanometer resolution. According to their mechanical response, the molecular mechanism of their biological functions can be unveiled [20]. The mechanical work done by the force or the torque can also be calculated and a quantitative free energy landscape can be obtained from the measured dynamic transition processes [21].

Popular single-molecule manipulation techniques include optical tweezers, atomic force microscopy (AFM), and magnetic tweezers (Figure 5.1). Optical tweezers, invented by Ashkin in the 1970s, were originally developed to trap micron-sized beads or atoms by a focused laser [22–25]. Subsequently, this technique was used as a biophysical tool to trap and manipulate bacteria, organelles or cells [26–29]. In the 1990s, optical tweezers were further developed as a single-molecule tool, as demonstrated in studies of DNA mechanics and molecular motors [30–35]. AFM was originally invented in the 1980s as a scanning probe microscopy technique to image biomolecules deposited on a flat surface [18,36,37]. In the 1990s, AFM was further developed as a single-molecule force spectroscopy tool and was applied to study force-dependent conformational transitions of biomolecules, such as protein unfolding [15, 38–40] and ligand–receptor interactions [41].

The concept of magnetic tweezers first appeared in the 1990s. A prototype of magnetic tweezers, developed by Bustamante et al., reported the first force–extension curve of a double-stranded DNA (dsDNA) [42, 43]. Later in 1996, the technique was further developed by Strick et al. to study the supercoiling properties of dsDNA, acknowledging its unique feature to apply a torque on the target analyte [44, 45]. DNA processing enzymes like helicases [46] and topoisomerases [47, 48] were also studied by magnetic tweezers.

In principle, the cantilever in an AFM or the laser trap in the optical tweezers acts as an elastic spring applying a stretching force to the biomolecule of interest. In such a scenario, force is correlated sensitively with the extension of the target biomolecule [12]. Consequently, any mechanical drift will significantly affect both the force and the extension during the measurement, limiting the stability of both systems in continuous, long-term measurements. To compensate for this, strict constant temperature and vibration-free environmental conditions have normally been required. In contrast, force applied on a paramagnetic bead using magnetic tweezers is controlled by a magnetic field with a gradient and thus the force applied is not sensitive to subtle distance changes on the scale of a molecule and is intrinsically constant. Extension measurements in magnetic tweezers rely solely on microscopic imaging, and are completely independent of the force generation system. These properties have endowed magnetic tweezers with a superior stability, capable of measuring the same biomolecule over a time scale of hours to days [21, 49].

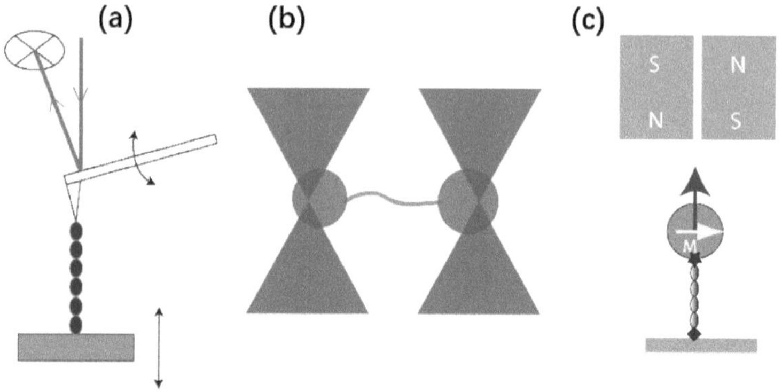

Figure 5.1 Schematics of single-molecule manipulation techniques: (a) AFM force spectroscopy, (b) optical tweezers, and (c) magnetic tweezers. Force generation and extension measurements are coupled in AFM and optical tweezers, while they are fully decoupled in magnetic tweezers, which makes magnetic tweezers the most stable technique of the three [50].

In this chapter, we will introduce the principles and technical details of magnetic tweezers, representative applications, and the emerging technical development. A detailed protocol is also presented, using as a demonstration of a highly simplified measurement of protein unfolding by magnetic tweezers.

5.2 Principles and Technical Details of Magnetic Tweezers

To build magnetic tweezers, two configurations such as vertical magnetic tweezers or transverse magnetic tweezers can be applied. The vertical magnetic tweezers were first developed by Strick et al. In this configuration, a stretching force perpendicular to the focal plane of an inverted microscope is applied [44]. However, with transverse magnetic tweezers, the target biomolecule is pulled in the focal plane of the microscope [51–53]. In this configuration, the biomolecule normally has a contour length of several micrometers. The vertical magnetic tweezers are more widely applied. Unless otherwise stated, all discussions here are based on the configuration of vertical magnetic tweezers.

To manipulate a single biomolecule using magnetic tweezers, the molecule of interest is tethered between a coverslip and a magnetic bead. The sample is placed on the stage of an inverted microscope equipped with manipulators, piezo objective actuators, and a high-speed camera. During the measurement, force and torque are applied to the molecule of interest via a tethered paramagnetic bead using an externally applied magnetic field and simultaneously, extension and twist are measured by microscopic imaging. To achieve a sub-pN force accuracy and a nanometer spatial resolution, a specially designed control system and image analysis algorithms are included. The physical principles and technical details related to force generation and calibration, torque control and extension measurement of magnetic tweezers are introduced below in more detail.

5.2.1 Force Generation

During measurements, a paramagnetic bead serves as a handle to apply the force and as a ruler to measure the extension of the molecule. Paramagnetic beads, such as Dynabeads, from Invitrogen Life Technologies, are widely used for this purpose. Experimentally, they can generate a force from less than 1 pN to more than 100 pN. The residual magnetic moment is almost negligible when the external magnetic field decreases to zero (Figure 5.2).

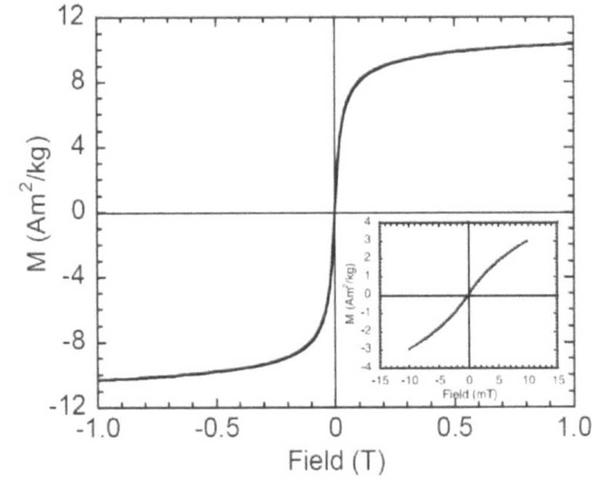

Figure 5.2 The magnetic moment of a Dynabeads M280 as a function of magnetic field without hysteresis, demonstrates that Dynabeads are ideal paramagnetic particles [54].

The potential energy of a paramagnetic bead in a magnetic field is given by the equation:

$$U = -\int_0^B \boldsymbol{M}(\boldsymbol{B}) \cdot d\boldsymbol{B} \tag{5.1}$$

where $\boldsymbol{B}$ is the magnetic field and $\boldsymbol{M}(\boldsymbol{B})$ is the induced magnetic moment (Figure 5.2). Normally the magnetic field $\boldsymbol{B}$ is strong enough that the magnetic moment has a saturated value $\boldsymbol{M}$. Thus, the potential energy can be simply described as $U \approx -\boldsymbol{M} \cdot \boldsymbol{B}$.

The stretching force $\boldsymbol{F}$ is equal to the negative gradient of energy U and is described as $\boldsymbol{F} = \nabla(\boldsymbol{M} \cdot \boldsymbol{B})$. Therefore, to apply a large force, the external magnetic field should be strong and should also possess a large spatial gradient. To achieve this, a pair of permanent magnetic rods or cubes is placed in an antiparallel orientation above the sample (Figure 5.3). Magnetic rods with a diameter of 3 to 4 mm or magnetic cubes with an edge length of 5 mm are commonly used [55, 56]. According to the force–distance function $F(d)$, the force F can be finely modulated by controlling the distance d between the permanent magnets and the sample. Usually $F(d)$ can be fitted by a single exponential function. However, when the magnets are very close to the sample, a double exponential function fits $F(d)$ better [55].

Electromagnets can also be used but the generated force amplitude is limited by the heating problem caused by the electromagnet. However, electromagnets have an advantage in the speed with which the stretching force can be modulated, and are suitable for applications requiring a high speed force modulation [57].

5.2.2 Force Calibration

Accurate measurement of the applied force is critical. Quantitatively, a paramagnetic bead tethered by a single molecule on a fixed point under a constant stretching force can be modeled as an inverted pendulum (Figure 5.3). According to this model, the restoring force f is related to the stretching force F: $f = -\frac{F}{l}\delta_y = -k\delta_y$, where l is the extension of the pendulum, and δ_y is the transverse deviation of the bead from its equilibrium position. $k = F/l$ is the effective spring constant of the potential well in the plane perpendicular to the stretching force. Therefore, from the equal partition theorem, the relationship between the stretching force and this transverse fluctuation can be derived as:

$$F = \frac{k_B T l}{\delta_y^2}, \tag{5.2}$$

where δ_y^2 is the variance of the transverse fluctuation [44]. A more general derivation of this equation has been published [58].

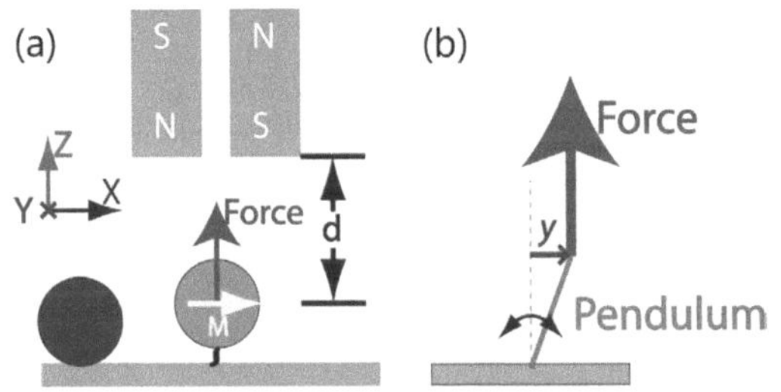

Figure 5.3 Model of an inverted pendulum for a tethered bead under constant upward stretching force. The effective length of the pendulum is the molecular extension plus the bead radius [55].

When two antiparallel magnetic rods are placed above the sample (Figure 5.3), the magnetic field at the position of the bead is perpendicular to the direction of the force which magnetizes the bead. More precisely, the pendulum composed of a tethered molecule and a paramagnetic bead has an effective extension of "$z+R$", where z is the extension of molecule and R is the bead radius [55] and the force calibration equation becomes:

$$F = \frac{k_B T (z+R)}{\delta_y^2}. \tag{5.3}$$

Extension of a molecule can be measured accurately by calculating the mean value of the measured extension over a long period of time. However, the variance of the transverse fluctuation δ_y^2 is affected by the blurring effect caused by the finite exposure time of the camera, which lasts for several milliseconds. Consequently, when the force is large and the tethered molecule is short, δ_y^2 may be under-estimated, causing the force to be over-estimated. The blurring effect can however be numerically corrected [59–62]:

$$\delta_{y,m}^2 = \delta_y^2 \left[\frac{2}{\alpha} - \frac{2}{\alpha^2} \left(1 - \exp(-\alpha) \right) \right] \tag{5.4}$$

Here, $\delta_{y,m}^2$ is the measured transverse variance, W is the exposure time of the camera, and α is the ratio of exposure time to the relaxation time of the bead: $\alpha = Wk/\gamma$. γ is the friction coefficient of the bead in the solution.

Magnetic forces applied to different paramagnetic beads usually lead to considerable variance. However, $F(d)$ from different paramagnetic beads can be universally described as $F(d) = C \times F^*(d)$, in which C is a bead-specific prefactor [55]. According to Stokes' law, $F^*(d)$ can be obtained by measuring the drifting velocity of the paramagnetic bead in viscous fluid (Figure 5.4). Specifically, a full range $F(d)$ of the bead being used can be determined by simple calibrations.

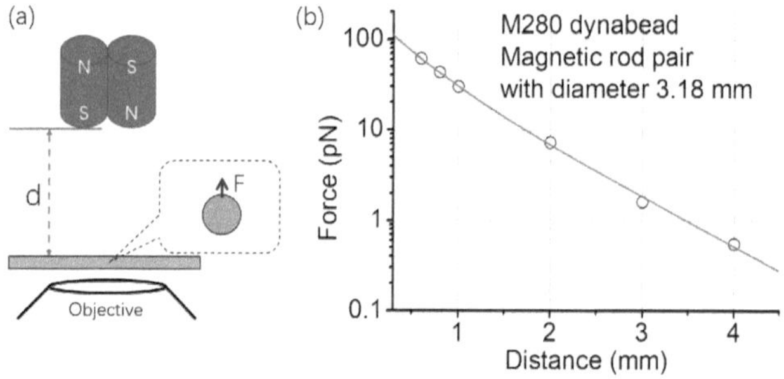

Figure 5.4 Force calibration of magnetic tweezers. (a) Sketch of magnetic tweezers with free paramagnetic beads M280 in 95% glycerol solution. (b) The magnetic force is calibrated by the drag force as a function of the distance between the magnetic rod pair and the sample channel.

5.2.3 Torque Generation and Measurement

If the bead is isotropically paramagnetic and without any residual magnetic moment, its magnetic moment will always be aligned with the external magnetic field. Thus, the net torque given by the equation $\tau = \boldsymbol{M} \times \boldsymbol{B}$ will always be zero. Commercially provided paramagnetic beads however have dispersed magnetic nanocrystals confined in small cavities of the polymer matrix, resulted in nonisotropic magnetic properties and a minor residual magnetic moment always exists [54]. Consequently, the magnetic moment of the bead $(\boldsymbol{M})$ is not perfectly aligned with the magnetic field $(\boldsymbol{B})$, generating a torque $\tau = \boldsymbol{M} \times \boldsymbol{B}$ which attempts to rotate the bead to reach an alignment between $\boldsymbol{M}$ and $\boldsymbol{B}$ [63]. This phenomenon has been widely used to twist biomolecules, such as dsDNA [44]. In this case, the magnetic field forms a torsional trap to constrain the orientation of a paramagnetic bead.

However, the angular deviation between $\boldsymbol{M}$ and $\boldsymbol{B}$ is too small to be accurately determined, and this causes difficulty in measurement of the torque applied to the paramagnetic bead. Lipfert and Dekker used a cylindrical magnet attached to a small magnet to generate a magnetic field with the main component along the force direction and a small component perpendicular to the force direction, forming a low-stiffness torsional trap. The elastic constant of this torsional trap can be determined from variance of the rotational angle of paramagnetic bead. The torque is equal to the deviation of bead twist angle from the magnetic field times the elastic constant of the torsional trap [64, 65]. Using such magnetic torque tweezers, the torsional stiffness of DNA and RecA-DNA filament was eventually measured [66].

5.2.4 Extension Measurements

The focal plane of the objective is defined as the *x–y* plane and the optical axis as the *z*-axis. The extension of a stretched biomolecule is measured from the three-dimensional location of a tethered bead. In the *x–y* plane, the bead position is determined from the centroid of its image and a nanometer spatial resolution can be achieved (Figure 5.5). The diffraction pattern of the bead out of focus appears as rings around the bead, whose pattern is sensitive to the out-of-focus distance. Therefore, analysis of the image pattern can help to determine the bead location in the *z* direction (Figure 5.6).

Usually a piezo objective actuator or a motorized microscope stage is used to adjust the focal plane with a nanometer resolution. Image patterns formed at different focal planes are stored to form an archive enabling image look up. Image comparison can be done in real space, in which the radius intensity profile from the bead centroid is obtained and the noise is reduced by azimuthal averaging [67]. Based on the intensity profile, an image pattern is compared with those prestored in the archive to estimate the z position of the bead. To achieve a nanometer resolution, correlation coefficients are calculated and quadratic interpolation is used to obtain the peak position of the correlation coefficient. Image comparison can also be done in the Fourier space. If the region of interest of a bead image is big enough to cover all diffraction rings of the bead, the z position can be derived from the comparison of the power spectrum of the bead image. The z position derived in this way is independent of the x and y centroid positions of the bead.

In practice, extension measurement depends on the configuration from which it is taken. With transverse magnetic tweezers, the tethered molecule is stretched in the focal plane of the microscope objective [51]. Extension of the molecule is directly measured by the bead centroid in the x–y plane. With vertical magnetic tweezers, force is applied along the optical axis. Extension of the molecule along the force direction is measured along the z-axis by image comparison.

$$x_{centroid} = \frac{\sum_{i,j} x_{i,j} I_{i,j}}{\sum_{i,j} I_{i,j}}$$

$$y_{centroid} = \frac{\sum_{i,j} y_{i,j} I_{i,j}}{\sum I_{i,j}}$$

Figure 5.5 The centroid of the bead image gives a nanometer spatial resolution of the bead location. Summation in the equations is over every pixel of the bead image region of interest. $I_{i,j}$ is the intensity of each pixel less the background intensity. The image of the bead is captured by a 100× oil immersion objective with back-scattered illumination. Under conditions of bright field illumination, the bead is dim while the background is bright. Therefore, the image intensity needs to be inverted before determining the centroid of the bead.

Noncoherent but approximately parallel light is used to illuminate the sample via either the bright field illumination [67] or the back-scattered illumination [55]. The bright field illumination requires a gap between the permanent magnets to form a light path. However, this configuration will decrease the maximum force that can be applied [56]. A benefit from bright field illumination is that the range of extension measurement can be more than 10 microns, great for the generation of a long-range force–extension curve. Back-scattered illumination uses the fluorescence illumination light path with a beam splitter instead of a dichromic mirror. Oil immersion objectives with large numerical apertures (NA) are usually used to image the bead. Clear ring patterns of the bead can be obtained in a focal range of 2 microns which is sufficient to study conformational transitions of short DNA or protein molecules (Figure 5.6).

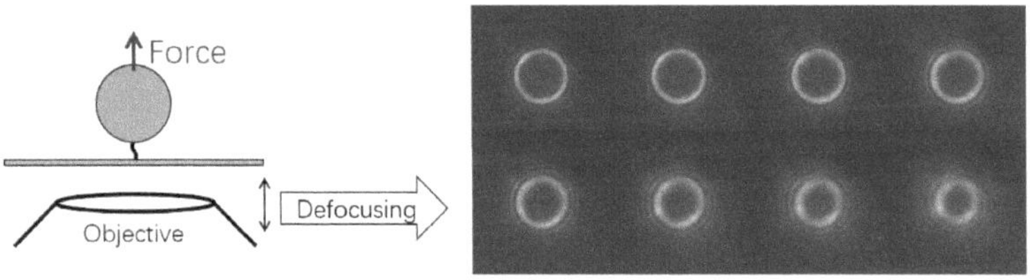

Figure 5.6 Images of a M280 Dynabead at different focal planes in a range of 2 microns. Olympus 100× NA 1.3 objective on IX71 inverted microscope is used to capture images with back-scattered illumination.

During long-term measurements, the sample will inevitably drift due to temperature fluctuations or mechanical vibrations from the environment. This drift can be passively compensated by using a fixed bead as a position reference. Alternatively, an active feedback system can be applied to stabilize the setup by a piezo objective actuator or a motorized stage by keeping the reference bead locked in the same position.

5.2.5 Data Analysis

Conformational transitions of biomolecules usually lead to abrupt jumps in the measured extension. Therefore, from a time course of the measured extension, stepwise extension jumps need to be detected and collected. Some data analysis algorithms were developed to automatically extract steps from the extension time courses [68–70]. Among these algorithms, hidden Markov modeling is widely used [71], has been further developed [72], and is universally applicable to analysis of other force spectroscopy results [73].

In summary, the force and the torque applied on the biomolecule are generated by controlling the magnitude, the gradient and the direction of the magnetic field applied on a magnetic bead. The extension or the twist angle is measured by analysis of the microscopic imaging results. Force–torque generation and extension–twist measurement in magnetic tweezers are fully decoupled, distinct from AFM or the optical tweezers. This has gained magnetic tweezers a superb stability and utility for long-term measurements.

Magnetic tweezers can apply a force from 0 pN to more than 100 pN, within the physiological force range. The extension can be measured with a nanometer spatial resolution and millisecond temporal resolution. The force response of biomolecules over a large force range, and conformation transition with rates over several orders of magnitude (from $<10^{-3}$ to 100 s^{-1}) can be studied with magnetic tweezers. Some slow dynamic processes with transition rates $<10^{-3}s^{-1}$, such as unfolding of mechanically stable proteins at small forces, can only be measured to date by magnetic tweezers [21, 74].

5.3 Applications of Magnetic Tweezers

Biomolecules that have been studied by magnetic tweezers include DNA, RNA, proteins, molecular motors, and many others. Most of the studies are related to mechanobiology-related problems. From the viewpoint of information flow, these problems can be categorized into two types: from a mechanical manipulation to a chemical response such as force or torque-induced conformation transitions and binding/unbinding interactions of DNA and proteins; and from a chemical reaction to a mechanical transformation such as directional translocation of molecular motors of DNA helicases using the energy from ATP hydrolysis. Below we describe some representative studies that have been performed with magnetic tweezers.

5.3.1 DNA Elasticity and Conformational Transition

Due to its long contour length, its wide availability and the ease of labeling of its termini, DNA is the first biomolecule to be studied by magnetic tweezers. Under physiological conditions, dsDNA is in the B-DNA form, a stable conformation with Watson–Crick base pairs. A variety of the mechanical properties of dsDNA, such as its bending and twist rigidity [42, 43], unzipping kinetics [75], and supercoiling conformation transitions [44, 76] can be directly studied with magnetic tweezers.

During the measurement, when the applied force is smaller than 20 pN, dsDNA was found to behave like a worm-like chain (WLC) with a bending persistence length of ~50 nm [42, 43]. With the ability to twist dsDNA using magnetic tweezers, it was found by Strick et al [44] that dsDNA can form a supercoiling structure when over- or under-twisted.

Beyond the critical force of 65 pN, the contour length of dsDNA increased by ~70%, but because the maximal force that could be applied by magnetic tweezers was limited at the early stages of its development, the famous DNA overstretching transition that happens at a ~65 pN stretching force was first discovered by optical tweezers [30] and optical fiber instrument [77]. However, an unexpected partial hysteresis phenomenon was observed when the force was decreased, which led to a long-term dispute about the nature of the DNA overstretching transition.

With the development of high-force magnetic tweezers which can generate stretching forces up to 100 pN, the DNA overstretching transition was thoroughly studied with different DNA sequences and under different environment conditions. Long dsDNA, with a contour length of several microns can be stretched by transverse magnetic tweezers, while short dsDNA composed of hundreds of base pairs can be stretched by vertical magnetic tweezers (Figure 5.7). Acknowledging the stability of a magnetic tweezers arrangement, temperature and salt concentration can be changed during measurement of the same DNA. This is, however, difficult for optical tweezers or AFM force spectroscopy. It was found that two distinct DNA overstretching transitions can happen under different environmental conditions: one without any hysteresis and one with a large hysteresis in the force-decreasing process [16, 17, 78]. Consequently, the nature of DNA overstretching transition depends on the detailed experimental conditions. Force-induced DNA peeling transition, which is similar to DNA melting with hysteresis occurs under experimental conditions such as high AT content, high temperature, and low salt concentration, which tend to destabilize dsDNA base pairing interaction. On the other hand, transition of B-DNA to S-DNA, a new conformation of dsDNA with longer contour length and helical structure [79], without hysteresis, occurs under conditions which stabilize DNA base pairs, such as high GC content, low temperature, and high salt concentration. Temperature-dependent study of the DNA overstretching transition allows direct measurement of the enthalpy and entropy changes during these two distinct transitions [78, 80].

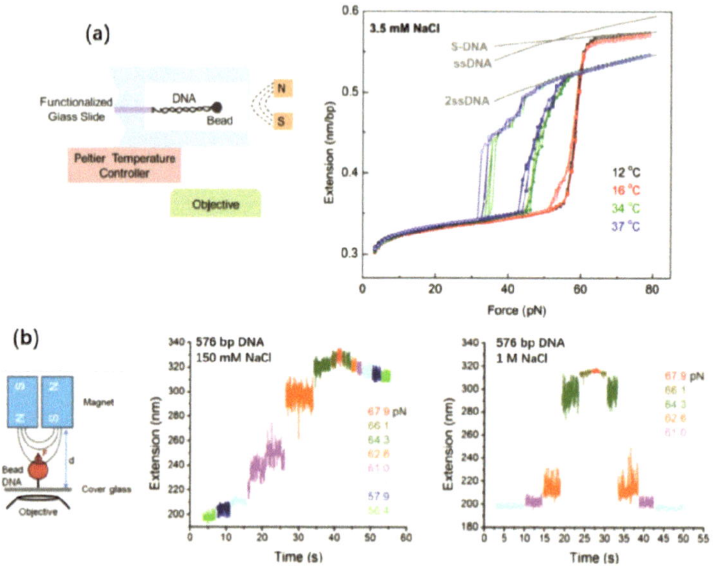

Figure 5.7 Mechanics of DNA overstretching transition at different temperatures and salt concentrations. (a) DNA overstretching transition is studied by transverse magnetic tweezers with temperature control, which reveals that dsDNA reverts to different conformations at different temperatures [78, 80]. (b) Short DNA of 576 base pairs is overstretched by vertical magnetic tweezers with nanometer resolution [17]. At 150 mM NaCl, the transition is not reversible, while at 1 M NaCl, the transition is reversible.

5.3.2 DNA Topoisomerase

DNA can form entangled conformations such as a supercoiled structure [81]. Natural DNA topoisomerase helps to disentangle the supercoiled structure of DNA during replication or transcription. There are two types of DNA topoisomerases: type I and type II. Type I cuts only one strand to release the supercoiling. Whereas type II cuts both strands [82, 83].

As magnetic tweezers can twist DNA to change its topological conformation [44], the activity of different types of topoisomerases can be studied by magnetic tweezers [47, 48, 84]. With a small stretching force, supercoiled DNA has a shorter extension than its relaxed form (Figure 5.8a). When a topoisomerase relaxes the DNA supercoiling, DNA extension reports an increase in its extension (Figure 5.8b) [85].

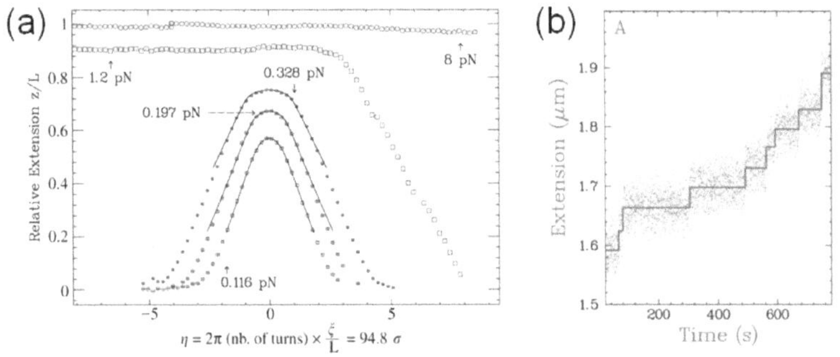

Figure 5.8 DNA supercoiling properties and mechanism of DNA topoisomerase. (a) Relative extension of the DNA double helix as a function of the degree of supercoiling, η, which was controlled by twisting of the paramagnetic bead at different constant forces of 8 pN, 1.2 pN, 0.328 pN, 0.197 pN, and 0.116 pN. The DNA extension decreases with increasing positive supercoiling at forces of 1.2 pN or less [44, 86]. (b) The supercoiling relaxation steps of positive supercoiled DNA by *Escherichia coli* topo I enzyme at a stretching force of 2 pN [48].

DNA gyrase, which belongs to topoisomerase II, can actively induce negative supercoiling using energy from ATP hydrolysis [87, 88]. Gyrase forms a complex with DNA and mechanically bends DNA of more than 100 bp to a specific conformation. Then it cuts both strands of the dsDNA to release the supercoiling. The whole process of conformation transition is coupled with steps of ATP binding, ATP hydrolysis, and ADP release. In a magnetic tweezers experiment, DNA can be labeled with an additional nonmagnetic bead as a tracker to measure both its extension and rotation [89–91], with which the working steps of DNA gyrase were observed in detail.

5.3.3 DNA and RNA Helicase

Helicases are a kind of molecular motor which can move along nucleic acid strands and separate complementary strands of a double-helix DNA or RNA either passively or actively by using the energy of nucleoside triphosphate (NTP) hydrolysis [92]. Re-annealing of separated ssDNA makes bulk experiments challenging. To avoid this problem, single-strand DNA binding proteins are frequently required during relevant studies in bulk. However, the confusion is much simplified in the single molecule regime [93, 94]. During single-molecule manipulation, the stretching force will inhibit the occurrence of re-annealing before dissociation of helicase (Figure 5.9). Because of the stable force control, magnetic tweezers have the advantage of performing long-term measurements to repetitively record events of helicase binding and unwinding on the same DNA construct [20, 94–97]. Additionally, the capability of temperature control also enables magnetic tweezers to study the temperature-dependent activity of helicase [98].

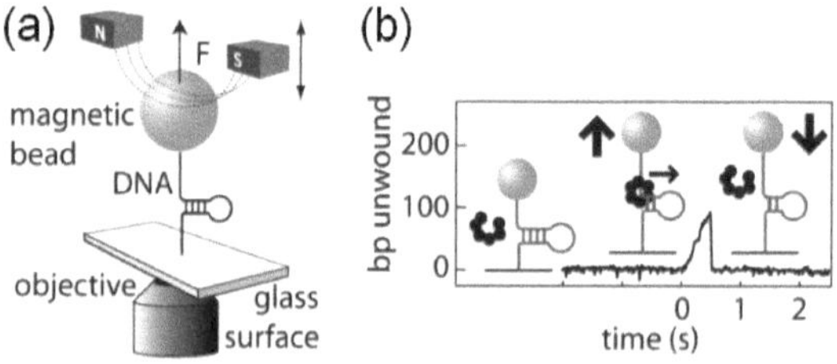

Figure 5.9 DNA unwinding by helicase gp41 measured by magnetic tweezers [96, 99]. (a) Sketch of DNA hairpin with 231 bp being stretched by a magnetic force smaller than the unzipping force of DNA with helicase gp41 protein in solution. (b) A slowly rising edge was caused by DNA unwinding and fast falling edge was due to the spontaneous DNA rehybridization upon gp41 dissociation.

There are different types of helicases which may work passively or actively. During magnetic tweezers experiments, the sequence-dependence and force-dependence of helicase activity reveal the working mode of different helicases [46]. In the passive mode, DNA strands in dsDNA split to ssDNA spontaneously due to thermal fluctuations. Before reannealing occurs, the helicase moves along the ssDNA directionally to the dsDNA–ssDNA fork to inhibit the re-hybridization of DNA. Consequently, the unwinding rate of passive helicase is sensitive to the DNA sequence and the stretching force which affect the thermal fluctuation of transient base pair opening. Though separation of DNA strands is due to thermal fluctuation, directional movement of the helicase also needs energy input from

NTP hydrolysis in the passive mode. In the active mode, separation of DNA strands is caused directly by the active translocation of helicase along the DNA using energy from NTP hydrolysis, and its unwinding rate is insensitive to the DNA sequence and the applied force.

5.3.4 DNA–Protein Interactions

In cells, DNA is well organized by DNA-binding proteins such as nucleoid association proteins in prokaryotic cells [18, 100] or histones in eukaryotic cells [101, 102]. Gene expression on the other hand is regulated by different kinds of transcription factors which may bind to the DNA promoter regions [103]. DNA-binding proteins can be classified based on their sequence selectivity into specific or nonspecific DNA-binding proteins. Nonspecific DNA-binding proteins bind to any DNA sequences with a similar affinity, while specific DNA-binding proteins preferentially bind to specific target sequences. Most DNA-binding proteins such as bacterial nucleoid association proteins HNS and IHF [18, 100] have properties in between these two categories, i.e., they bind to random sequences with a low affinity and bind to their target sequences with a high affinity.

A high concentration of nonspecific DNA-binding protein binds to DNA to form a protein–DNA complex. Properties of this protein–DNA complex, including changes of the contour length, the persistence length, the binding affinity, the binding dynamics and the interaction patterns can be well characterized by magnetic tweezers. Experimentally, a single unbound dsDNA is first identified by measuring its force–extension curve over a force range of 0.1 pN to 10 pN. The curve can be fit to a WLC model. Then the protein is introduced into the measurement chamber to trigger the formation of a protein–DNA complex. Force–extension curves of a protein–DNA complex are subsequently measured (Figure 5.10). Different force–extension curves indicate different binding modes of the protein on the DNA [104]. If the protein binds to DNA to form a rigid filament complex, the force–extension curve appears as a WLC response with a persistence length longer than that of the unbound form, such as DNA-binding protein HNS in solution without divalent ions [18]. If the protein bends DNA locally, the force–extension curve shows a shorter persistence length. This is exemplified by proteins HMGB1, NHP6A, and HU [105]. If the protein cross-links with remote binding sites along DNA, the DNA extension will collapse to a short length progressively when the force drops below a critical value, such as the protein HNS in solution with magnesium ions [18] or the protein FIS [106]). When a large stretching force is applied again the DNA extension will increase abruptly due to the opening of DNA loops.

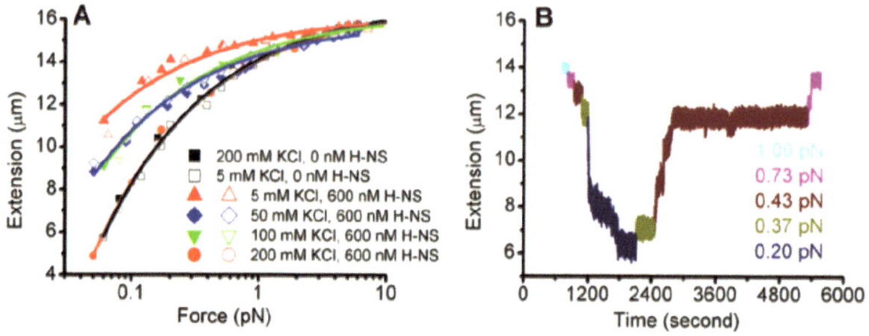

Figure 5.10 Interaction of protein HNS with dsDNA depends on the presence of divalent cations [18]. (a) Without divalent cations in solution, HNS will polymerize along the DNA backbone to form a rigid DNA–HNS complex with longer persistence length than naked dsDNA. (b) With 10 mM magnesium ions in solution, HNS crosslinks dsDNA, and causes DNA to collapse when the force is smaller than 0.2 pN. When force of 0.43 pN or bigger is applied, extension increases abruptly.

In a single molecular study of nonspecific DNA-binding proteins, although one DNA molecule may be stretched by the force, there are usually hundreds to thousands of proteins simultaneously interacting with the same DNA at random potential binding sites. The binding event of a single protein is hard to detect because of the intrinsic extension fluctuation of the long DNA molecule.

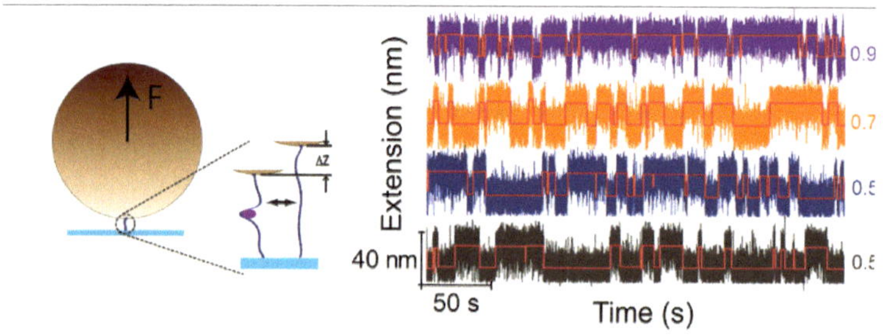

Figure 5.11 Specific interaction between IHF and its binding site on dsDNA [107]. After binding, DNA is kinked to decrease its extension. Therefore, real-time association and dissociation dynamics of IHF can be studied from the extension time trace by magnetic tweezers.

To detect the signal of single-protein binding, short DNA can be stretched in a vertical magnetic tweezers experiment to improve its spatial resolution due to less extension fluctuation of short DNA. The protein IHF binds to its specific binding site and bends DNA locally. Because locally curved DNA has a shorter extension than the straight B-form DNA (Figure 5.11), association and dissociation of a single IHF protein on DNA were detected in real time by detection of the extension decreasing and increasing steps [107].

5.3.5 Protein Folding and Unfolding

A DNA template to study specific protein–DNA interactions can be around 100 nm in length, while the size of a single folded protein measures only several nanometers. Unlike optical tweezers which have interference problems when two tethered beads are too close to each other, magnetic tweezers are free from this issue. A protein can be directly tethered between a magnetic bead and the substrate, without a need for excessively long linker molecules. Thus, effect of protein linker conjugation is conserved in studies by magnetic tweezers instead of optical tweezers.

Protein tags can be used to link proteins between a coverslip and the paramagnetic bead. However, most protein tags were developed for protein purifications, which don't require a strong binding affinity [108]. To minimize the chance of the tether breaking during manipulation, other tags such as HaloTag [109] or SpyTag [110], which form covalent bonds with their binding targets, were developed. Biotin can be ligated to AviTag and form a strong binding with streptavidin [108] or traptavidin [111] coated on paramagnetic beads.

Dynamics of both protein unfolding and folding can be studied with magnetic tweezers. With a protein sandwiched between a coverslip and a paramagnetic bead, an unfolding event of a protein domain can be detected by a sudden increase of its extension at a certain stretching force. When the applied force decreases below a critical value, the unfolded protein can refold again, registering a sudden decrease of its extension length.

Force-dependent unfolding and folding can be studied with a constant loading rate, in which the force increases or decreases linearly with time [108, 112]. The unfolding or folding force distribution that is obtained depends on the loading rate. This force distribution and its dependence on the loading rate can be analyzed to obtain useful dynamic parameters of the unfolding or folding process [113, 114].

Alternatively, protein unfolding and folding processes can be studied by constant force measurements [21, 74]. Magnetic tweezers can intrinsically apply a constant force and are suitable for measurements of this mode. In contrast, to maintain a constant stretching force, AFM force spectroscopy requires a complicated feedback servo control and may report complex folding dynamics, possibly due to the interference of the servo control system [115]. However with magnetic tweezers, which intrinsically apply a constant force, directly reporting stepwise folding and unfolding of proteins, consistent with the classical two-state dynamics of protein folding [21]. This demonstrates the importance of equipment stability, and reminds us that explaining the data from complicated setups with a feedback control can be tricky.

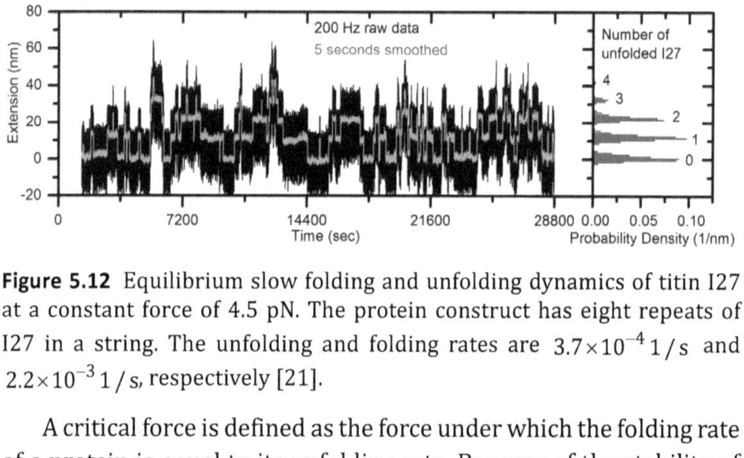

Figure 5.12 Equilibrium slow folding and unfolding dynamics of titin I27 at a constant force of 4.5 pN. The protein construct has eight repeats of I27 in a string. The unfolding and folding rates are 3.7×10^{-4} $1/s$ and 2.2×10^{-3} $1/s$, respectively [21].

A critical force is defined as the force under which the folding rate of a protein is equal to its unfolding rate. Because of the stability of the magnetic tweezers setup, the equilibrium folding and unfolding process of proteins with very slow folding and unfolding rates can be recorded by magnetic tweezers (Figure 5.12). For example, immunoglobular domain I27 from the muscle protein, titin, folds and unfolds with a transition rate of 3×10^{-4} s^{-1} under its critical force of 5.4 pN [21]. AFM usually can unfold proteins at high forces from tens of pN to hundreds of pN, but the physiologically relevant mechanical response of proteins to low forces usually cannot be measured by AFM. Direct measurement of the protein folding and unfolding rates at small forces by magnetic tweezers can reveal complex protein dynamics. For example, we discovered an unexpected catch bond-like behavior of I27 unfolding at forces smaller than 20 pN [74].

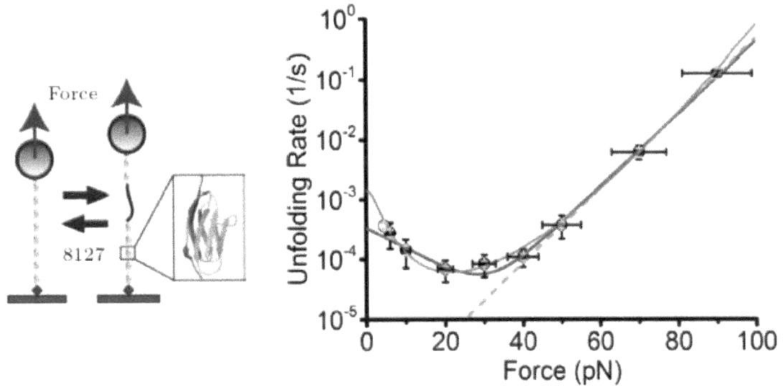

Figure 5.13 Catch-bond behavior of titin I27 unfolding dynamics. At a force smaller than 20 pN, the stretching force makes I27 unfold more slowly with increasing force [74].

5.3.6 Protein–Protein Interactions

Protein–protein interactions play a crucial role in biological signaling networks. Cytoskeleton networks in the cell, adhesion complexes and the extracellular matrix are all composed of protein complexes. The formation of a protein complex with two components can be simplified as the formula: $S+P \leftrightarrow SP$, where S and P are two proteins while SP is the formed complex. If the protein complex helps to crosslink cytoskeletons, cellular membranes and extracellular matrices, a mechanical force which comes from molecular motor inside the cell or mechanical stimuli outside the cell will be sustained. Therefore, the force-dependent interaction of these proteins is crucial to their biological functions, such as the mechanical sensing of cells [1, 4].

To study the mechanical stability of a protein complex, force can be applied to pull apart the interacting protein pair. But after that, the tether will break, prohibiting efficient acquisition of data to form the statistics. A smart design is to use a flexible polymer as a linker to connect both interacting proteins. Therefore, after force-induced dissociation, the tether does not break and the proteins will associate together to form the protein complex again after force relaxation [116, 117].

Then multiple cycles of experiments can be done with the same tether to produce the statistics.

To study the force-dependent protein–protein interactions, a protein S can be pulled by a force, while a protein P is freely placed in the solution with varying concentrations. If binding of P modulates the mechanical stability of S, its force response will have a concentration dependence of P (Figure 5.14). Sometimes the binding interaction also depends on the pulling force which changes the conformation of protein S. For example, a force will partially unfold the protein talin to expose its vinculin binding site. In this way, it was found that a force in the range of 5 pN to 20 pN will enhance the binding of vinculin to talin by several orders of magnitude [118]. Similar to talin, the interaction between α-catenin and vinculin has similar properties [119].

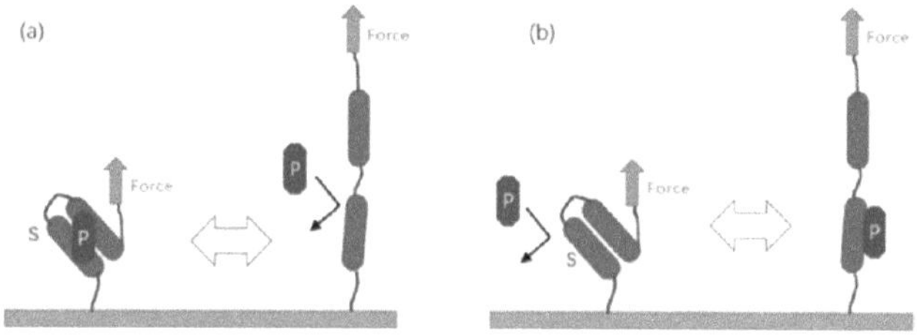

Figure 5.14 A principle with which to study force-dependent protein–protein interactions. (a) If stable native structure of substrate protein "S" is necessary to bind protein "P", force-induced unfolding will cause dissociation. (b) If binding sites of "P" are hidden inside native structure of "S", force-induced unfolding of "S" will expose the binding sites and facilitate binding of protein "P". On the other hand, protein binding will modulate the folding and unfolding dynamics of the substrate protein.

5.3.7 Mechanical Manipulation of Cells

Magnetic tweezers can also be used to manipulate cells. To stimulate cells with a local force using magnetic tweezers, paramagnetic or ferromagnetic beads are attached to the cell surface. In contrast to a single molecule, a much larger force, on the order of several nano-Newtons, is usually required to obtain an observable response from cells. Therefore, the magnet used in these studies is normally in a tip geometry and is placed extremely close to the bead [120].

Torque can also be applied to the cell via ferromagnetic beads. Torque constraint applies a mechanical load on the cell locally without a significant perturbation of the shape of the cell. With this strategy, Wang et al. coated Arg-Gly-Asp (RGD) ligands on a bead and found that integrins work as mechanoreceptors to transmit external mechanical stimuli into cells and cause reorganization of the cytoskeleton [121]. Further work found that RGD or lysine-coated beads produce different mechanical anisotropic responses in living cells, which revealed that cell tension through stress fibers causes the mechanical anisotropy of cells [122, 123]. Deng et al. measured the viscous-elastic property of the cytoskeleton through repeated torque stimuli to cells, and found that the fast response of the cytoskeleton originates from entropic elasticity of semiflexible polymers, and the slow dynamics at time scales longer than 10 milliseconds arises from the soft glass properties of the cytoskeleton [124].

5.4 Emerging Developments

Since the invention of magnetic tweezers, the core technique has been continuously developed to meet the emerging need from single-molecule studies. Below we introduce several representative types of specially designed magnetic tweezers.

5.4.1 Freely-Orbiting Magnetic Tweezers

Conventional magnetic tweezers use two antiparallel magnetic rods to generate a strong magnetic field with a large gradient. The magnetic field and the induced magnetic moment are both perpendicular to the force direction and consequently, the magnetic bead cannot rotate freely around the symmetric axis. Lipfert and Dekker developed freely-orbiting magnetic tweezers using a single magnetic cylinder to generate a magnetic field whose gradient is along the field direction, and with which the bead can freely rotate around the symmetric axis. If a tethered dsDNA attaches to the bead and the glass surface with both strands, then a rotation of the bead will report the twist of dsDNA [125]. When a single dsDNA molecule is tethered to a paramagnetic bead on a glass surface, the tether point is usually not at the bottom pole of the bead. The centroid of the bead will form a circle in the x–y plane when dsDNA twists around its axis. Therefore, the three-dimensional track of the bead centroid gives both the extension and the twist angle of the dsDNA (Figure 5.15).

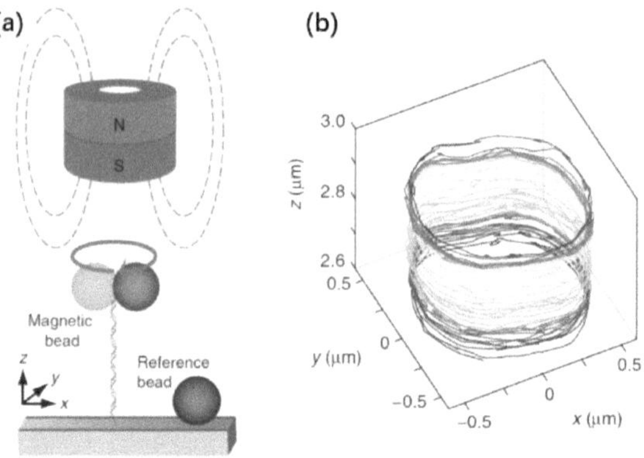

Figure 5.15 Freely-orbiting magnetic tweezers. (a) Sketch of freely-orbiting magnetic tweezers recording the twisting of dsDNA. (b) When RecA proteins bind to dsDNA to form an elongated RecA/DNA filament complex, both the extension and the twist of the dsDNA double helix are recorded simultaneously [125].

5.4.2 Combination of Magnetic Tweezers with Fluorescence

During single-molecule manipulation measurements, extension of the molecule is usually the only detectable parameter to report its conformational transition. To detect the association or dissociation of its binding partner, or to get extra sensing information from another dimension, fluorescence detection can also be incorporated into magnetic tweezers.

Optical tweezers have been combined with fluorescence imaging to determine the conformation of dsDNA after overstretching transitions [126]. However, local heating caused by the trapping laser might also affect the transition pathways [127]. The strongly focused trapping laser also significantly interferes with fluorescence imaging by causing photobleaching [128, 129]. Different from optical tweezers, force or torque generated in magnetic tweezers is modulated by the external magnetic field, which does not interfere at all with the fluorescence excitation and emission spectrum. This is convenient for simultaneous measurement of fluorescence imaging.

To reach the sensitivity to resolve single-molecule fluorescence, total internal reflection fluorescence (TIRF) illumination is combined with vertical magnetic tweezers. The paramagnetic bead is far away from the TIRF illumination layer and autofluorescence of the bead doesn't interfere with the fluorescence signal from target molecules [130, 131]. With this strategy, it was found that force-dependent interactions between vinculin and talin play an important role in the mechano-sensing processes of cells [132].

Similarly, gold particle labels and micro-mirror TIRFs were applied to achieve simultaneous measurement of twist, torque, and extension of DNA [133]. Because gold particles cannot be photo-bleached and have a well-defined size, they work as an ideal reporter of both DNA extension and twist. Simultaneously, the drag force during rotation accurately reports the torque. As reported, multidimensional measurements (force vs extension or torque vs twist) revealed the detailed molecular mechanism of DNA gyrase by this system [133].

5.4.3 Fast Dynamics Studied by Electromagnets

The slow response of the motorized stage has significantly restrained the speed with which the applied force can be modulated using magnetic tweezers constructed with permanent magnets. This impedes the capability to capture fast dynamics of target biomolecules. Recently, Fernandez et al. developed a magnetic tweezers arrangement with an electromagnet-based head, with which a local magnetic field with a high gradient can be generated [57]. In this setup, a force with up to 50 pN can be applied on a Dynabead M270 bead by operating with a 1 Ampère current through the electromagnet. The force can be modulated every 0.1 millisecond, acknowledging the small inductance of the magnetic head and the wide bandwidth of the controlling circuit. Working with a high-speed camera, fast dynamic processes with a submillisecond temporal resolution can now be studied.

5.5 Summary and Perspectives

In summary, magnetic tweezers have demonstrated a great capability to manipulate a single biomolecule or a cell with a well-controlled stretching force, torque or a combination thereof. Biological samples have negligible magnetic susceptibility, indicating that an externally applied magnetic field would not interact with most biological samples. The magnetic force or torque applied via a magnetic bead is thus not interfered by the biomolecule or the cell being studied. Different from that of optical tweezers or AFM force spectroscopy, the force generation in magnetic tweezers is completely decoupled from the extension measurement system, making magnetic tweezers much more stable for continuous long-term measurements. Thus currently, direct measurement of slow dynamic processes can only be performed by magnetic tweezers.

Magnetic tweezers have been widely used to study polymer elasticity, molecular motors, conformational transitions, and interactions of biomolecules. With the advances in new camera and illumination technologies, the temporal and spatial resolutions of magnetic tweezers have been respectively improved to the submillisecond and the subnanometer regime. The stretching force of magnetic tweezers is highly controllable within the physiological force range, making the measured force response of biomolecules directly related to their biological functions in vivo.

Appendix: A Tutorial Protocol: Observing Protein Folding/Unfolding Using Magnetic Tweezers

Purpose

The free energy of protein folding determines its structural stability, while the rates of folding/unfolding transitions reflect the pathway and the energy barrier of the transitions. Magnetic tweezers, which apply a well-controlled force to the protein termini, are able to directly monitor the folding and unfolding dynamics of a single protein through its extension time courses. The streptococcal B1 immunoglobulin-binding domain of protein G (GB1), which possesses an exceptional elastomeric feature and is mechanically stable under tension, was tentatively selected as the model protein for this study. Its polyprotein form, consisting of eight identical tandem repeats of GB1 domains and desired tethers, was engineered, and we demonstrate in detail how the target protein is conjugated and how its mechanical property is measured.

Materials

- **Equipment**
  Oven (Senxin, China)
  Ultrasonic cleaner (Zhiwei, China)
  Plasma cleaner (Diener, Germany)
  Hotplate (Zinique Lab)
  Inverted microscope (Eclipse Ti-U, Nikon, Japan)
  Motorized translation stage (PI, Germany)
  Piezo objective actuator (PI, Germany)
  CMOS camera (Basler, Germany)
  Floating anti-vibration optical table (Liansheng, China)
  Desktop computer with customized software (Dell, US)
- **Consumables**
  Glass coverslips (Thermo Fisher)
  Parafilm (Sigma Aldrich)
  Silicone Glue (3140 RTV, Dow Corning)
  Polybead Amino Microspheres 3.00 μm (Polysciences)
  Dynabead M280 (Invitrogen Life Technologies)
- **Reagents**
  Deionized water (Milli-Q)
  Decon 90 (Decon, UK)
  PBS (Sangon Biotech, China)
  3-Aminopropyltriethoxysilane (APTES), (Sigma Aldrich)
  Sulfosuccinimidyl4-(N-maleimidomethyl) cyclohexane-1-carboxylate (Sulfo-SMCC), (ThermoFisher Scientific)
  Bovine serum albumin (BSA)(Lee Biosolutions)
  E. coli strain BL21 (New England Biolab)
  Luria-Bertani (LB) medium (Sangon Biotech, China)
  Restriction enzymes (New England Biolab)
  Isopropylthio-β-galactoside (IPTG),(Sangon Biotech, China)
  Antibiotics ampicillin and chloramphenicol (Sangon Biotech, China)

Safety and Precautions

- APTES can cause severe skin burns and eye damage. Wear appropriate gloves and lab coat and handle it in a fume hood.

- Coverslips could easily break and may have sharp edges. Wear appropriate gloves and goggles. Dispose used coverslips into a container for sharp objects.

Methods

Step 1: Adjustment of Magnetic Tweezers

The magnetic tweezers arrangement is based on an inverted microscope (Eclipse Ti-U, Nikon, Japan) equipped with a 4× air objective and a 100× oil-immersion objective. A piezo objective actuator (PI, Germany) is used to finely adjust the focal plane of the objective. A motorized linear translation stage (PI, Germany) is used to adjust the distance between the permanent magnets and the sample. A collimated LED lamp (Thorlabs, USA) is installed at the back port of the microscope to illuminate the sample (Figure 5.16a). Both the piezo objective actuator and the motorized stage are controlled by a custom LabVIEW program. Microscopic images are captured and analyzed in real time to get the three-dimensional location of the paramagnetic bead. Calibration of the setup must be performed prior to the measurement.

1.1 Magnetic rods alignment: Mount the magnetic rods on a motorized linear translation stage. Place the translation stage above the microscope stage. Mount a 4× air objective on the piezo objective actuator. Move the magnetic rods toward the objective. Finely adjust the focus so that the bottom surface of the magnetic rods is imaged. Finely adjust the *x*–*y* position of magnetic rods using the translation stage so that the symmetric center of the rods is aligned with the optical axis (Figure 5.16b).

1.2 Zero-distance calibration: Move up the magnetic rods. Place one piece of 1.5# 24 mm × 40 mm coverslip on the microscope stage. Slowly move down the magnetic rods toward the coverslip. Stop the motor when the magnets are just in contact with the coverslip. Then set this position as the zero-distance position.

Step 2: Sample Preparation

2.1 Gene cloning and protein expression: The plasmid coding for eight repeats of GB1 protein with desired termini tags (Histag-Avitag-$GB1_8$-Spytag) (Figure 5.17a) is prepared by gene cloning.

The plasmid, which is ampicillin-resistant, is transformed into *E. coli* BL21 host cells. Another plasmid, which is chloramphenicol resistant and express biotin ligase BirA, is simultaneously transformed into the same host cells. The cells are cultured at 37°C with vigorous stirring (200 rpm) in a LB medium containing 100 mg/L of ampicillin and 34 mg/L of chloramphenicol. When the optical density (OD) at 600 nm of the bacterial suspension reaches 0.6 to 0.8, protein expression is induced with IPTG at a 1 mM final concentration. Subsequently, the culture is incubated at 37°C for further 4 hours to produce the desired protein which is in vivo biotinylated by the co-expressed BirA enzyme. The protein is harvested and purified with cobalt-nitrilotriacetic acid (Co-NTA) resin, and stored in a 1× PBS buffer. The purified protein is aliquoted to 10 μL per fraction and stored at -80°C.

2.2 **Cleaning of coverslips:** Place #1.5 coverslips with a size of 22 mm × 40 mm and 22 mm × 22 mm in two separate staining jars. Fill both jars with an aqueous solution of 1% (v/v) Decon 90 so that the coverslips are fully immersed. Place the jars in an ultrasonic cleaner and clean for 10 minutes. Discard the Decon 90 solution, rinse the coverslips with deionized (DI) water for 20 times and dry them in a 100°C oven for 20 minutes. Afterward, treat the coverslips with oxygen plasma for 10 minutes.

2.3 **APTES coating:** Place the 22 mm × 40 mm coverslips in a staining jar. Fill the staining jar with 1% (v/v) methanol solution of APTES so that the coverslips are fully immersed. Incubate it in a fume hood for 60 minutes. Afterward, discard the methanol solution and wash it sequentially with methanol and deionized water. Dry the coverslip in a 100°C oven for 1 hour.

2.4 **Flow channel assembly:** Prepare a piece of parafilm, measuring 22 mm × 40 mm. Cut a rectangular hole, measuring about 15 mm × 30 mm in the parafilm. The flow channel is formed by sandwiching the above mentioned parafilm between an APTES-coated 22 mm × 40 mm coverslip and a 22 mm × 22 mm coverslip. Apply silicone glue to the interface of the channel to form a sealed reservoir. The silicone glue is set overnight prior to use (Figure 5.16c).

2.5 **Reference bead deposition:** Dilute the Polybead Amino Microsphere stock solution 200 times with 1× PBS. Re-suspend and wash it with 1× PBS. Flow the diluted bead solution into the flow channel and incubate for 20 minutes. A majority of beads will stick to the bottom surface of the channel due to nonspecific adhesion. Free beads are eluted with 1× PBS out of the channel. It is desired to have 2 to 3 stuck microspheres as reference beads in each field of view during microscopic imaging.

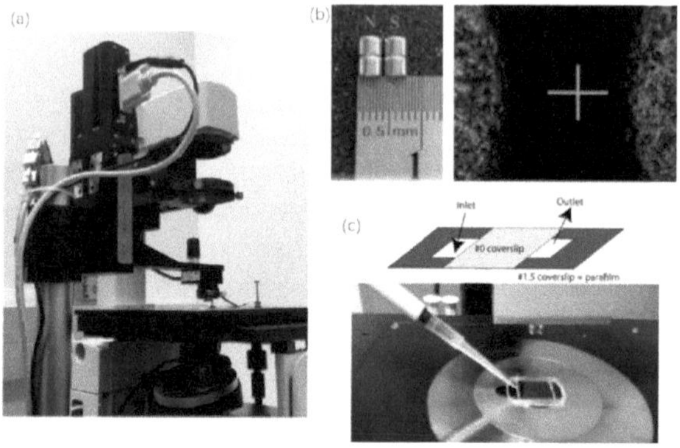

Figure 5.16 A magnetic tweezers setup and the flow channel. (a) Magnetic tweezers are built on an inverted microscope, following a vertical configuration. (b) Permanent magnetic rods and their microscopic image captured by a 4× air objective. The magnetic rods are aligned with the center of the image (marked by a green cross) to ensure that an upward pulling force is applied. (c) The flow channel is made from coverslips and a parafilm clip. The picture shows a prepared flow channel on a microscope stage. The analyte is loaded by pipetting the solution to the inlet of the channel.

2.6 **SpyCatcher protein coating:** 1× PBS solution of Sulfo-SMCC at a ~1 mg/mL concentration is flowed into the channel and incubated for 20 minutes. Subsequently, 1× PBS solution of SpyCatcher protein with a concentration of ~10 ng/μL is flowed into the channel and incubated for 2 hours. Afterward, wash the channel with 200 μL 1× PBS. Subsequently, fill the channel with 10 mg/mL BSA dissolved in a Tris buffer (Tris 20 mM, pH 7.4, NaCl 150 mM) and incubate overnight to passivate the glass surface.

2.7 **Tethering a paramagnetic bead to the protein of interest:** The purified protein of Histag-Avitag-GB1$_8$-Spytag is diluted to 1 nanomolar (nM) with 1× PBS. Hundred microliter diluted solution is flowed into the SpyCatcher-coated channel followed with a 10 minute incubation. Dilute the stock solution of M280 Dynabead with 1× PBS buffer. Flow 30 μL of diluted Dynabead into the channel and incubate for ~5 minutes. Excessive, untethered M280 beads are removed by flowing the channel with 1× PBS buffer. Beads left in the chamber are conjugated to the bottom surface of the chamber, in a configuration described in Figure 5.17a. This flow chamber is now ready for subsequent measurements.

Step 3: Magnetic Tweezers Measurements

Place the prepared flow chamber on the microscope stage. Finely adjust the focus to image the beads on the bottom surface of the channel. Move down the magnets to a distance of ~4 mm above the zero distance position (corresponding to a force of less than 1 pN). The tethered paramagnetic beads can be recognized from their characteristic motion fluctuations in the *x*–*y* plane. The bead with a single-molecule tether tends to have larger fluctuations. Slide the microscope stage so that the bead of interest is moved to the center of view. A stuck bead nearby can be used as a reference. Adjust the focus to make the focal plane slightly below the best focusing position, then start to capture images at a series of focal planes. Store these images at different focal planes into an image library to form a look-up table. Then start the measurement.

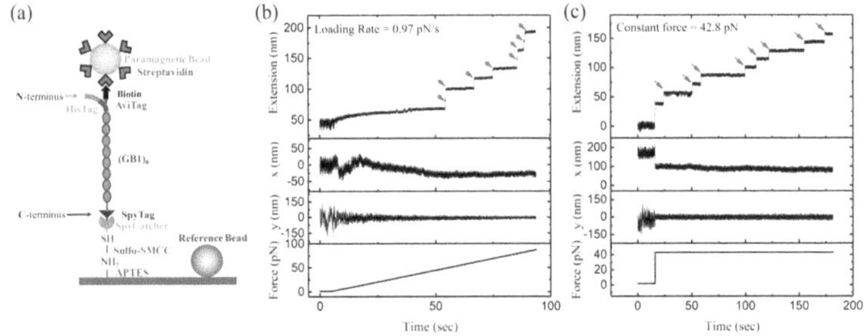

Figure 5.17 Unfolding dynamics of GB1 protein acquired by magnetic tweezers. (a) Sketch of the protein construct Histag-Avitag-GB1$_8$-Spytag during a magnetic tweezers measurement. (b) A representative time course containing unfolding steps of GB1, acquired by the constant loading rate mode of measurement. The trace was recorded with a constant loading rate of 0.97 pN/s. (c) A representative time course containing unfolding steps of GB1, acquired by the force jump mode of measurement. The applied force was kept at 42.8 pN. Note that in time course of *x* and *y*, there is no sudden jump when GB1 domains were unfolded, which indicates that there is only one single protein construct linking the paramagnetic bead to the coverslip surface.

3.1 Constant loading rate measurement: By moving the magnets toward or away from the sample, the force applied on the bead can be modulated. The applied force can be gradually increased with time following the equation $F = rt$, where r is the loading rate, forming the constant loading rate mode of measurement. Simultaneously, the extension of the molecule is recorded. During the acquired time course, a sudden stepwise jump of extension can be observed, indicating unfolding of a protein GB1 domain (Figure 5.17b). By rising up the applied force, all eight GB1 domains were sequentially unfolded when the applied force is above 50 pN. The distribution of the force under which a GB1 protein was unfolded reflects the mechanical stability of GB1.

3.2 **Force jump measurement:** Unfolding rates at a constant stretching force can be measured by the force jump method. When a force of ~1 pN is applied, all eight GB1 domains are in their native states. Then the force is immediately increased to 42.8 pN. At this constant stretching force, extension increases stepwise due to unfolding of each individual GB1 domain (Figure 5.17c). Force-dependent unfolding rates can be obtained from such force jump measurement with different stretching forces.

References

1. Moore SW, Roca-Cusachs P, Sheetz MP (2010). Stretchy proteins on stretchy substrates: the important elements of integrin-mediated rigidity sensing, *Dev Cell,* **19**, 194–206.
2. Vogel V, Sheetz M (2006). Local force and geometry sensing regulate cell functions, *Nat Rev Mol Cell Biol,* **7**, 265–275.
3. Vilar JM, Saiz L (2005). DNA looping in gene regulation: from the assembly of macromolecular complexes to the control of transcriptional noise, *Curr Opin Genet Dev,* **15**, 136–144.
4. Vogel V (2006). Mechanotransduction involving multimodular proteins: converting force into biochemical signals, *Annu Rev Biophys Biomol Struct,* **35**, 459–488.
5. Ciobanasu C, Faivre B, Le Clainche C (2013). Integrating actin dynamics, mechanotransduction and integrin activation: the multiple functions of actin binding proteins in focal adhesions, *Eur J Cell Biol,* **92**, 339–348.
6. Parsons JT, Horwitz AR, Schwartz MA (2010). Cell adhesion: integrating cytoskeletal dynamics and cellular tension, *Nat Rev Mol Cell Biol,* **11**, 633–643.
7. Fletcher DA, Mullins RD (2010). Cell mechanics and the cytoskeleton, *Nature,* **463**, 485–492.
8. Heisenberg CP, Fassler R (2012). Cell-cell adhesion and extracellular matrix: diversity counts, *Curr Opin Cell Biol,* **24**, 559–561.

9. Engler AJ, Sen S, Sweeney HL, Discher DE (2006). Matrix elasticity directs stem cell lineage specification, *Cell,* **126**, 677–689.
10. Panorchan P, et al. (2006). Single-molecule analysis of cadherin-mediated cell-cell adhesion, *J Cell Sci,* **119**, 66–74.
11. Hinterdorfer P, Oijen Av (2009). *Handbook of Single-Molecule Biophysics*, Springer.
12. Neuman KC, Nagy A (2008). Single-molecule force spectroscopy: optical tweezers, magnetic tweezers and atomic force microscopy, *Nat Methods,* **5**, 491–505.
13. Dulin D, Lipfert J, Moolman MC, Dekker NH (2013). Studying genomic processes at the single-molecule level: introducing the tools and applications, *Nat Rev Genet,* **14**, 9–22.
14. Leake MC (2017). *Biophysics: Tools and Techniques*, CRC Press.
15. Puchner EM, Gaub HE (2009). Force and function: probing proteins with AFM-based force spectroscopy, *Curr Opin Struct Biol,* **19**, 605–614.
16. Fu H, Chen H, Marko JF, Yan J (2010). Two distinct overstretched DNA states, *Nucleic Acids Res,* **38**, 5594–5600.
17. Fu H, et al. (2011). Transition dynamics and selection of the distinct S-DNA and strand unpeeling modes of double helix overstretching, *Nucleic Acids Res,* **39**, 3473–3481.
18. Liu Y, Chen H, Kenney LJ, Yan J (2010). A divalent switch drives H-NS/DNA-binding conformations between stiffening and bridging modes, *Genes Dev,* **24**, 339–344.
19. van Oene MM, et al. (2017). Applying torque to the *Escherichia coli* flagellar motor using magnetic tweezers, *Sci Rep,* **7**, 43285.
20. Seol Y, Strub MP, Neuman KC (2016). Single molecule measurements of DNA helicase activity with magnetic tweezers and t-test based step-finding analysis, *Methods,* **105**, 119–127.
21. Chen H, et al. (2015). Dynamics of equilibrium folding and unfolding transitions of titin immunoglobulin domain under constant forces, *J Am Chem Soc,* **137**, 3540–3546.
22. Ashkin A (1970). Acceleration and trapping of particles by radiation pressure, *Phys Rev Lett,* **24**, 156.
23. Chu S, Hollberg L, Bjorkholm JE, Cable A, Ashkin A (1985). Three-dimensional viscous confinement and cooling of atoms by resonance radiation pressure, *Phys Rev Lett,* **55**, 48–51.
24. Chu S, Bjorkholm JE, Ashkin A, Cable A (1986). Experimental observation of optically trapped atoms, *Phys Rev Lett,* **57**, 314–317.
25. Ashkin A (1987). Laser manipulation of atoms, *Nature,* **330**, 608–609.
26. Ashkin A, Dziedzic JM (1987). Optical trapping and manipulation of viruses and bacteria, *Science,* **235**, 1517–1520.

27. Ashkin A, Dziedzic JM (1989). Internal cell manipulation using infrared laser traps, *Proc Natl Acad Sci USA,* **86**, 7914–7918.
28. Ashkin A, Schutze K, Dziedzic JM, Euteneuer U, Schliwa M (1990). Force generation of organelle transport measured in vivo by an infrared laser trap, *Nature,* **348**, 346–348.
29. Ashkin A, Dziedzic JM, Yamane T (1987). Optical trapping and manipulation of single cells using infrared laser beams, *Nature,* **330**, 769–771.
30. Smith SB, Cui Y, Bustamante C (1996). Overstretching B-DNA: the elastic response of individual double-stranded and single-stranded DNA molecules, *Science,* **271**, 795–799.
31. Block SM, Goldstein LS, Schnapp BJ (1990). Bead movement by single kinesin molecules studied with optical tweezers, *Nature,* **348**, 348–352.
32. Svoboda K, Schmidt CF, Schnapp BJ, Block SM (1993). Direct observation of kinesin stepping by optical trapping interferometry, *Nature,* **365**, 721–727.
33. Svoboda K, Block SM (1994). Force and velocity measured for single kinesin molecules, *Cell,* **77**, 773–784.
34. Block SM (1995). Macromolecular physiology. One small step for myosin, *Nature,* **378**, 132–133.
35. Yin H, et al. (1995). Transcription against an applied force, *Science,* **270**, 1653–1657.
36. Binnig G, Quate CF, Gerber C (1986). Atomic force microscope, *Phys Rev Lett,* **56**, 930–933.
37. Bao Q, et al. (2007). A divalent metal-mediated switch controlling protein-induced DNA bending, *J Mol Biol,* **367**, 731–740.
38. Fisher TE, Oberhauser AF, Carrion-Vazquez M, Marszalek PE, Fernandez JM (1999). The study of protein mechanics with the atomic force microscope, *Trends Biochem Sci,* **24**, 379–384.
39. Alessandrini A, Facci P (2005). AFM: a versatile tool in biophysics, *Meas Sci Technol,* **16**, R65–R92.
40. Rief M, Gautel M, Oesterhelt F, Fernandez JM, Gaub HE (1997). Reversible unfolding of individual titin immunoglobulin domains by AFM, *Science,* **276**, 1109–1112.
41. Florin EL, Moy VT, Gaub HE (1994). Adhesion forces between individual ligand-receptor pairs, *Science,* **264**, 415–417.
42. Smith SB, Finzi L, Bustamante C (1992). Direct mechanical measurements of the elasticity of single DNA molecules by using magnetic beads, *Science,* **258**, 1122–1126.

43. Bustamante C, Marko JF, Siggia ED, Smith S (1994). Entropic elasticity of lambda-phage DNA, *Science,* **265**, 1599–1600.
44. Strick TR, Allemand JF, Bensimon D, Bensimon A, Croquette V (1996). The elasticity of a single supercoiled DNA molecule, *Science,* **271**, 1835–1837.
45. Mosconi F, Allemand JF, Bensimon D, Croquette V (2009). Measurement of the torque on a single stretched and twisted DNA using magnetic tweezers, *Phys Rev Lett,* **102**, 078301.
46. Manosas M, Xi XG, Bensimon D, Croquette V (2010). Active and passive mechanisms of helicases, *Nucleic Acids Res,* **38**, 5518–5526.
47. Charvin G, Bensimon D, Croquette V (2003). Single-molecule study of DNA unlinking by eukaryotic and prokaryotic type-II topoisomerases, *Proc Natl Acad Sci USA,* **100**, 9820–9825.
48. Dekker NH, et al. (2002). The mechanism of type IA topoisomerases, *Proc Natl Acad Sci USA,* **99**, 12126–12131.
49. Popa I, et al. (2016). A HaloTag anchored ruler for week-long studies of protein dynamics, *J Am Chem Soc,* **138**, 10546–10553.
50. Qian H, Chen H, Yan J (2016). Frontier of soft matter experimental technique: single molecular manipulation, *Acta Phys Sinica,* **65**, 188706.
51. Yan J, Skoko D, Marko JF (2004). Near-field-magnetic-tweezer manipulation of single DNA molecules, *Phys Rev E,* **70**, 011905.
52. Liu YJ, Chen H, Kenney LJ, Yan J (2010). A divalent switch drives H-NS/DNA-binding conformations between stiffening and bridging modes, *Genes Dev,* **24**, 339–344.
53. Yu M, et al. (2017). mDia1 senses both force and torque during F-actin filament polymerization, *Nat Commun,* **8**, 1650.
54. Fonnum G, Johansson C, Molteberg A, Morup S, Aksnes E (2005). Characterisation of Dynabeads (R) by magnetization measurements and Mossbauer spectroscopy, *J Magn Magn Mater,* **293**, 41–47.
55. Chen H, et al. (2011). Improved high-force magnetic tweezers for stretching and refolding of proteins and short DNA, *Biophys J,* **100**, 517–523.
56. Lipfert J, Hao X, Dekker NH (2009). Quantitative modeling and optimization of magnetic tweezers, *Biophys J,* **96**, 5040–5049.
57. Tapia-Rojo R, Eckels EC, Fernandez JM (2019). Ephemeral states in protein folding under force captured with a magnetic tweezers design, *Proc Natl Acad Sci USA,* **116**, 7873–7878.
58. Yan J, Kawamura R, Marko JF (2005). Statistics of loop formation along double helix DNAs, *Phys Rev E,* **71**, 061905.

59. Wong WP, Halvorsen K (2006). The effect of integration time on fluctuation measurements: calibrating an optical trap in the presence of motion blur, *Opt Express*, **14**, 12517–12531.
60. Klaue D, Seidel R (2009). Torsional stiffness of single superparamagnetic microspheres in an external magnetic field, *Phys Rev Lett*, **102**, 028302.
61. te Velthuis AJ, Kerssemakers JW, Lipfert J, Dekker NH (2010). Quantitative guidelines for force calibration through spectral analysis of magnetic tweezers data, *Biophys J*, **99**, 1292–1302.
62. Yu Z, et al. (2014). A force calibration standard for magnetic tweezers, *Rev Sci Instrum*, **85**, 123114.
63. van Oene MM, et al. (2015). Biological magnetometry: torque on superparamagnetic beads in magnetic fields, *Phys Rev Lett*, **114**, 218301.
64. Lipfert J, Kerssemakers JW, Jager T, Dekker NH (2010). Magnetic torque tweezers: measuring torsional stiffness in DNA and RecA-DNA filaments, *Nat Methods*, **7**, 977–980.
65. Lipfert J, Lee M, Ordu O, Kerssemakers JWJ, Dekker NH (2014). Magnetic tweezers for the measurement of twist and torque, *J Vis Exp*, **87**, 51503.
66. Lipfert J, et al. (2011). Magnetic torque tweezers and their application in probing the torsional properties of DNA, RNA, and DNA filaments, *Biophys J*, **100**, 176–176.
67. Gosse C, Croquette V (2002). Magnetic tweezers: micromanipulation and force measurement at the molecular level, *Biophys J*, **82**, 3314–3329.
68. Carter BC, Vershinin M, Gross SP (2008). A comparison of step-detection methods: how well can you do?, *Biophys J*, **94**, 306–319.
69. Little MA, et al. (2011). Steps and bumps: precision extraction of discrete states of molecular machines, *Biophys J*, **101**, 477–485.
70. Arunajadai SG, Cheng W (2013). Step detection in single-molecule real time trajectories embedded in correlated noise, *PLoS One*, **8**, e59279.
71. Stigler J, Rief M (2012). Hidden Markov analysis of trajectories in single-molecule experiments and the effects of missed events, *Chemphyschem*, **13**, 1079–1086.
72. Zhang Y, Jiao J, Rebane AA (2016). Hidden Markov modeling with detailed balance and its application to single protein folding, *Biophys J*, **111**, 2110–2124.
73. Stigler J, Ziegler F, Gieseke A, Gebhardt JC, Rief M (2011). The complex folding network of single calmodulin molecules, *Science*, **334**, 512–516.

74. Yuan G, et al. (2017). Elasticity of the transition state leading to an unexpected mechanical stabilization of titin immunoglobulin domains, *Angew Chem Int Ed Engl,* **56**, 5490–5493.

75. Koch SJ, Wang MD (2003). Dynamic force spectroscopy of protein-DNA interactions by unzipping DNA, *Phys Rev Lett,* **91**, 028103.

76. Sheinin MY, Wang MD (2009). Twist-stretch coupling and phase transition during DNA supercoiling, *Phys Chem Chem Phys,* **11**, 4800–4803.

77. Cluzel P, et al. (1996). DNA: an extensible molecule, *Science,* **271**, 792–794.

78. Zhang X, Chen H, Fu H, Doyle PS, Yan J (2012). Two distinct overstretched DNA structures revealed by single-molecule thermodynamics measurements, *Proc Natl Acad Sci USA,* **109**, 8103–8108.

79. Leger JF, et al. (1999). Structural transitions of a twisted and stretched DNA molecule, *Phys Rev Lett,* **83**, 1066–1069.

80. Zhang X, et al. (2013). Revealing the competition between peeled ssDNA, melting bubbles, and S-DNA during DNA overstretching by single-molecule calorimetry, *Proc Natl Acad. Sci USA,* **110**, 3865–3870.

81. Fuller FB (1978). Decomposition of the linking number of a closed ribbon: a problem from molecular biology, *Proc Natl Acad Sci USA,* **75**, 3557–3561.

82. Wang JC (1996). DNA topoisomerases, *Annu Rev Biochem,* **65**, 635–692.

83. Seol Y, Neuman KC (2011). Single-molecule measurements of topoisomerase activity with magnetic tweezers, *Methods Mol Biol,* **778**, 229–241.

84. Koster DA, Croquette V, Dekker C, Shuman S, Dekker NH (2005). Friction and torque govern the relaxation of DNA supercoils by eukaryotic topoisomerase IB, *Nature,* **434**, 671–674.

85. Gunn KH, Marko JF, Mondragon A (2018). Single-molecule magnetic tweezer analysis of topoisomerases, *Methods Mol Biol,* **1703**, 139–152.

86. Strick T, Allemand J-F, Croquette V, Bensimon D (2000). Twisting and stretching single DNA molecules, *Prog Biophys Mol Biol,* **74**, 115–140.

87. Nollmann M, et al. (2007). Multiple modes of *Escherichia coli* DNA gyrase activity revealed by force and torque, *Nat Struct Mol Biol,* **14**, 264–271.

88. Ogawa T, et al. (2015). Direct observation of DNA overwinding by reverse gyrase, *Proc Natl Acad Sci USA,* **112**, 7495–7500.

89. Gore J, et al. (2006). Mechanochemical analysis of DNA gyrase using rotor bead tracking, *Nature,* **439**, 100–104.

90. Basu A, Schoeffler AJ, Berger JM, Bryant Z (2012). ATP binding controls distinct structural transitions of *Escherichia coli* DNA gyrase in complex with DNA, *Nat Struct Mol Biol,* **19**, 538–546, S531.
91. Basu A, Parente AC, Bryant Z (2016). Structural dynamics and mechanochemical coupling in DNA gyrase, *J Mol Biol,* **428**, 1833–1845.
92. Caruthers JM, McKay DB (2002). Helicase structure and mechanism, *Curr Opin Struct Biol,* **12**, 123–133.
93. Sun B, Wang MD (2016). Single-molecule perspectives on helicase mechanisms and functions, *Crit Rev Biochem Mol Biol,* **51**, 15–25.
94. Hodeib S, et al. (2017). A mechanistic study of helicases with magnetic traps, *Protein Sci,* **26**, 1314–1336.
95. Fiorini F, Bagchi D, Le Hir H, Croquette V (2015). Human Upf1 is a highly processive RNA helicase and translocase with RNP remodelling activities, *Nat Commun,* **6**, 7581.
96. Li JH, et al. (2016). Pif1 is a force-regulated helicase, *Nucleic Acids Res,* **44**, 4330–4339.
97. Raj S, et al. (2019). Mechanistic characterization of the DEAD-box RNA helicase Ded1 from yeast as revealed by a novel technique using single-molecule magnetic tweezers, *Nucleic Acids Res,* **47**, 3699–3710.
98. Gollnick B, et al. (2015). Probing DNA helicase kinetics with temperature-controlled magnetic tweezers, *Small,* **11**, 1273–1284.
99. Lionnet T, Spiering MM, Benkovic SJ, Bensimon D, Croquette V (2007). Real-time observation of bacteriophage T4 gp41 helicase reveals an unwinding mechanism, *Proc Natl Acad Sci USA,* **104**, 19790–19795.
100. Lin J, Chen H, Droge P, Yan J (2012). Physical organization of DNA by multiple non-specific DNA-binding modes of integration host factor (IHF), *PLoS One,* **7**, e49885.
101. Xiao X, et al. (2018). Dissection of structural dynamics of chromatin fibers by single-molecule magnetic tweezers, *Biophys Rep,* **4**, 222–232.
102. Li W, et al. (2016). FACT remodels the tetranucleosomal unit of chromatin fibers for gene transcription, *Mol Cell,* **64**, 120–133.
103. Browning DF, Grainger DC, Busby SJ (2010). Effects of nucleoid-associated proteins on bacterial chromosome structure and gene expression, *Curr Opin Microbiol,* **13**, 773–780.
104. Yan J, Marko JF (2003). Effects of DNA-distorting proteins on DNA elastic response, *Phys Rev E,* **68**, 011905.
105. Skoko D, Wong B, Johnson RC, Marko JF (2004). Micromechanical analysis of the binding of DNA-bending proteins HMGB1, NHP6A, and HU reveals their ability to form highly stable DNA-protein complexes, *Biochemistry,* **43**, 13867–13874.

106. Skoko D, Yan J, Johnson RC, Marko JF (2005). Low-force DNA condensation and discontinuous high-force decondensation reveal a loop-stabilizing function of the protein Fis, *Phys Rev Lett,* **95**, 208101.

107. Le S, et al. (2013). Mechanosensing of DNA bending in a single specific protein-DNA complex, *Sci Rep,* **3**, 3508.

108. Chen H, et al. (2011). Differential mechanical stability of filamin A rod segments, *Biophys J,* **101**, 1231–1237.

109. Taniguchi Y, Kawakami M (2010). Application of HaloTag protein to covalent immobilization of recombinant proteins for single molecule force spectroscopy, *Langmuir,* **26**, 10433–10436.

110. Zakeri B, et al. (2012). Peptide tag forming a rapid covalent bond to a protein, through engineering a bacterial adhesin, *Proc Natl Acad Sci USA,* **109**, E690–E697.

111. Chivers CE, et al. (2010). A streptavidin variant with slower biotin dissociation and increased mechanostability, *Nat Methods,* **7**, 391–393.

112. Chen H, et al. (2013). Mechanical perturbation of filamin A immunoglobulin repeats 20-21 reveals potential non-equilibrium mechanochemical partner binding function, *Sci Rep,* **3**, 1642.

113. Dudko OK, Hummer G, Szabo A (2008) Theory, analysis, and interpretation of single-molecule force spectroscopy experiments, *Proc Natl Acad Sci USA,* **105**, 15755–15760.

114. Pierse CA, Dudko OK (2017). Distinguishing signatures of multipathway conformational transitions, *Phys Rev Lett,* **118**, 088101.

115. Fernandez JM, Li HB (2004). Force-clamp spectroscopy monitors the folding trajectory of a single protein, *Science,* **303**, 1674–1678.

116. Kim J, Zhang CZ, Zhang X, Springer TA (2010). A mechanically stabilized receptor-ligand flex-bond important in the vasculature, *Nature,* **466**, 992–995.

117. Le S, Yu M, Yan J (2019). Phosphorylation reduces the mechanical stability of the alpha-catenin/ beta-catenin complex, *Angew Chem Int Ed Engl,* **58**, 18663–18669.

118. Yao M, et al. (2014). Mechanical activation of vinculin binding to talin locks talin in an unfolded conformation, *Sci Rep,* **4**, 4610.

119. Yao M, et al. (2014). Force-dependent conformational switch of alpha-catenin controls vinculin binding, *Nat Commun,* **5**, 4525.

120. Tanase M, Biais N, Sheetz M (2007). Magnetic tweezers in cell biology, *Methods Cell Biol,* **83**, 473–493.

121. Wang N, Butler JP, Ingber DE (1993). Mechanotransduction across the cell surface and through the cytoskeleton, *Science,* **260**, 1124–1127.

122. Hu S, et al. (2004). Mechanical anisotropy of adherent cells probed by a three-dimensional magnetic twisting device, *Am J Physiol Cell Physiol,* **287**, C1184–C1191.
123. Zhang Y, et al. (2017). Interfacing 3D magnetic twisting cytometry with confocal fluorescence microscopy to image force responses in living cells, *Nat Protoc,* **12**, 1437–1450.
124. Deng L, et al. (2006). Fast and slow dynamics of the cytoskeleton, *Nat Mater,* **5**, 636–640.
125. Lipfert J, Wiggin M, Kerssemakers JW, Pedaci F, Dekker NH (2011). Freely orbiting magnetic tweezers to directly monitor changes in the twist of nucleic acids, *Nat Commun,* **2**, 439.
126. van Mameren J, et al. (2009). Unraveling the structure of DNA during overstretching by using multicolor, single-molecule fluorescence imaging, *Proc Natl Acad Sci USA,* **106**, 18231–18236.
127. Peterman EJ, Gittes F, Schmidt CF (2003). Laser-induced heating in optical traps, *Biophys J,* **84**, 1308–1316.
128. Comstock MJ, Ha T, Chemla YR (2011). Ultrahigh-resolution optical trap with single-fluorophore sensitivity, *Nat Methods,* **8**, 335–340.
129. Dijk MA, Kapitein LC, Mameren J, Schmidt CF, Peterman EJ (2004). Combining optical trapping and single-molecule fluorescence spectroscopy: enhanced photobleaching of fluorophores, *J Phys Chem B,* **108**, 6479–6484.
130. Lee M, Kim SH, Hong SC (2010). Minute negative superhelicity is sufficient to induce the B-Z transition in the presence of low tension, *Proc Natl Acad Sci USA,* **107**, 4985–4990.
131. Long X, Parks JW, Stone MD (2016). Integrated magnetic tweezers and single-molecule FRET for investigating the mechanical properties of nucleic acid, *Methods,* **105**, 16–25.
132. del Rio A, et al. (2009). Stretching single talin rod molecules activates vinculin binding, *Science,* **323**, 638–641.
133. Lebel P, Basu A, Oberstrass FC, Tretter EM, Bryant Z (2014). Gold rotor bead tracking for high-speed measurements of DNA twist, torque and extension, *Nat Methods,* **11**, 456–462.

Chapter 6

Long-Time Recording of Single-Molecule Dynamics in Solution by Anti-Brownian Trapping

Quan Wang, Elif Karasu, and Hugh Wilson
Lewis-Sigler Institute for Integrative Genomics, Princeton University, Princeton, New Jersey 08544, USA
quanw@princeton.edu

Single-molecule fluorescence-based techniques have emerged to provide powerful and versatile means of sensing nanoscale structure and dynamics. In aqueous solution, molecular diffusion limits the typical observation time of a single molecule to ~1 ms, posing a bottleneck to the amount of information that can be extracted. To resolve this fundamental limitation, the Anti-Brownian ELectrokinetic (ABEL) trap was developed and enables long-term (~10 s), information-rich interrogation of single molecules in solution. In this chapter, we review the basic principles and technical development of the ABEL trap, as well as recent applications of the device to dissect dynamic processes at the nanoscale level. Finally, we provide a tutorial protocol that illustrates the procedures key to an ABEL trap experiment.

Single-Molecule Tools for Bioanalysis
Edited by Shuo Huang

ISBN 978-981-4800-44-0 (Hardcover), 978-1-003-18913-8 (eBook)
www.jennystanford.com

6.1 Introduction

Over the past 20 years, single-molecule fluorescence has been developed into a powerful suite of techniques in many disciplines of science (Moerner, Shechtman, and Wang 2015). In particular, single molecules are nanoscale sensors which can unravel both structural and functional insights of biological systems. To probe structural information (e.g., subcellular protein organization), single molecules can be used as nanometer light sources to sequentially light-up biological structures with a spatial resolution of ~20 nm, about one-tenth of the diffraction limit. This particular application of single-molecule fluorescence imaging, also known as "localization-based super-resolution microscopy" (Betzig et al. 2006; Hess, Girirajan, and Mason 2006; Rust, Bates, and Zhuang 2006), was recognized as part of the 2014 Nobel Prize in Chemistry (for recent reviews, see (Hell et al. 2015; Liu, Lavis, and Betzig 2015; Sahl and Moerner 2013)).

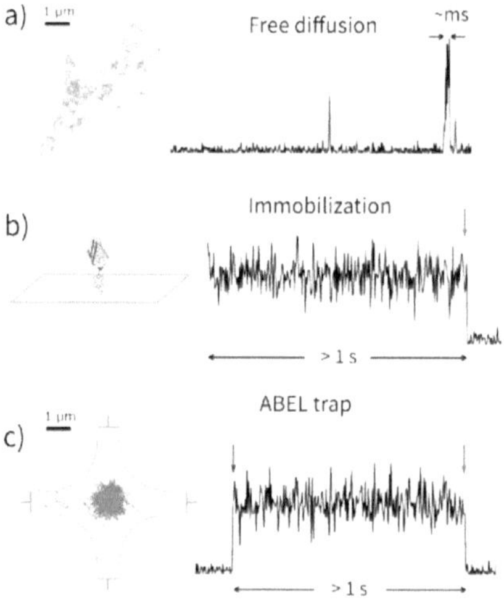

Figure 6.1 Three single-molecule measurement strategies. (a) Confocal detection in solution: molecules are freely diffusing but the observation window is limited to ~1ms due to Brownian motion. (b) Engineering a surface anchor on the single molecule enables long observation times by eliminating Brownian motion, but could perturb the biomolecule of interest. (c) The ABEL trap enables long observation times on freely diffusing molecules by active feedback suppression of Brownian motion. Blue arrow indicates a single-molecule entrance event. Red arrows in (b) and (c) indicate photobleaching events. Modified with permission from Wang, PhD Thesis, Stanford University, 2014.

Single-molecule fluorescence techniques have also been developed to provide functional and mechanistic details of biomolecules. These include, but are not limited to, single-molecule Forster resonance energy transfer (smFRET) (Ha et al. 1996; Joo et al. 2008; Schuler 2013; Sustarsic and Kapanidis 2015), fluorescence correlation spectroscopy (FCS) (Ries and Schwille 2012; Rigler and Elson 2012), multicolor colocalization microscopy (Boehm et al. 2016; Friedman, Chung, and Gelles 2006; Larson, Rodgers, and Hoskins 2014) and many others. Detailed mechanisms of DNA processing (Ha, Kozlov, and Lohman 2012), protein assembly (Hoskins et al. 2011), conformation dynamics of proteins (Yang et al. 2003), molecular motors (Yildiz 2003), catalysis (English et al. 2006), and numerous other biomolecular processes have been revealed by these techniques. Despite the enormous progress in these areas, two fundamental challenges still remain: a measurement strategy in a biologically compatible environment with minimum perturbation, and extracting the maximum amount of information from individual molecules.

An aqueous solution is a biologically compatible environment in which to probe biomolecules. However, in order to reach the desired signal-to-background ratio, a tightly focused confocal configuration is often used (Nie, Chiu, and Zare 1994), and the molecule in an aqueous solution easily escapes from the diffraction-limited focal volume (~1 fL) due to Brownian motion (caused by the perpetual thermal agitation from surrounding solvent molecules). As a consequence, the observation time of a single molecule is limited to about 1 ms (Figure 6.1a), and is insufficiently long to observe processes that take place on biological time scales. Various methods have been developed over the years to enable single-molecule observation in solution for longer times. The most widely used method is by surface immobilization or encapsulation into nano-containers (Figure 6.1b). Although effective and well-developed (Chandradoss et al. 2014; Hua et al. 2014), immobilization faces the unavoidable question of whether attachment to a surface of molecules would cause a perturbation to their intrinsic function. In many studies, the effect of the surface was minimal (Rasnik, McKinney, and Ha 2005), but in a few others, it was significant (Butler 2000; Friedel, Baumketner, and Shea 2006; Liu et al. 2013). In the context of biomolecular interactions, surface immobilized molecules have reduced the access volume and have restricted rotational degrees of freedom (i.e., caused a shift in the interaction free energy) compared to those free in solution, which might lead to altered binding rates compared to those measured by bulk studies.

Encapsulation into nano-containers is another popular strategy to extend observation times of single molecules in solution. This has been accomplished using lipid vesicles (Boukobza, Sonnenfeld, and Haran 2001; Cisse et al. 2007; Keller et al. 2012), hydrogel (Rahmanseresht et al. 2015), nano-holes (Rissin and Walt 2006; Rondelez et al. 2005; Shon and Cohen 2012) or engineered DNA nanostructures (Zhao et al. 2016). Nevertheless, nano-entrapment has a number of disadvantages, including a low yield (of capturing one molecule per container) and the difficulty of buffer exchanges and undesired nano-confinement effects (Rubinovich and Polak 2013) which originates from altered configuration entropy.

Another strategy which can enable long observation time is to "trap" objects in an aqueous solution using an external force field. Perhaps the most well-known approach in this category is optical tweezers (Ashkin et al. 1986; Moffitt et al. 2008). However, optical tweezers could not directly trap nanometer-sized biomolecules without the assistance of microscale beads. This is because the restoring forces in an optical trap become diminishingly small for nanometer sized objects ($F \propto r^3$, with r the radius of the object). Consequently, specially engineered local fields have to be implemented to trap objects smaller than 100 nm (Berthelot et al. 2014; Chen et al. 2012; Pang and Gordon 2012; Yang et al. 2009). Since the polarizability-induced optical forces are too weak to manipulate nanometer-sized biomolecules, other forces, such as electrostatic (Carlson et al. 2010; Krishnan et al. 2010), thermophoretic (Braun and Cichos 2013), and dielectrophoretic (Kuzyk et al. 2008) forces, have been exploited for trapping. In particular, electrokinetic forces, which result from the direct action of an external electric field on charged molecules (including both the target and its surrounding solvent molecules), can be used to manipulate nano-objects in solution, and this forms the basis of the anti-Brownian Electrokinetic (ABEL) trap technology (Figure 6.1c), the focus of this chapter.

6.2 Principles of Anti-Brownian Trapping

6.2.1 Fundamentals

The basic idea of the ABEL trap (Cohen and Moerner 2005) is the following: as a single molecule randomly diffuses in solution, its Brownian displacements are compensated by feedback electrokinetic forces in order to keep the molecule near the center of the field of view. In other words, if thermal motion drives the molecule to the right (with respect to a predefined "trapping center"), the feedback apparatus apply an electric field in solution to move the molecule back to the left, and vice versa. This feedback cycle is repeated thousands of times per second to lock the target molecule to the center (Figure 6.2). Effectively, this process creates a "virtual" potential well (Cohen 2005; Jun and Bechhoefer 2012) for the single molecule of interest, with a strength much stronger than that of optical tweezers (Moerner 2007). As a result, the ABEL trap is able to trap objects that are much smaller than those that are trappable in an optical tweezers.

6.2.2 A Brief History of the Development of the Technique

Several lines of technological advancement inspired and enabled the ABEL trap. First, single-molecule fluorescence imaging (Moerner 1999) provided a direct means to detect individual biomolecules of interest and track their positions in real time. Second, feedback control has been widely used to stabilize a variety of processes and biological entities. Rich theoretical and practical results were established to implement an optimal feedback loop. It is interesting to note that, as early as 1971, Berg (Berg 1971) developed a special microscope to track the motion of individual bacteria using feedback control. Third, capillary electrophoresis has been a mature analytical technique for separating and characterizing different molecular species by their distinct responses to electric fields (Dovichi and Zhang 2000).

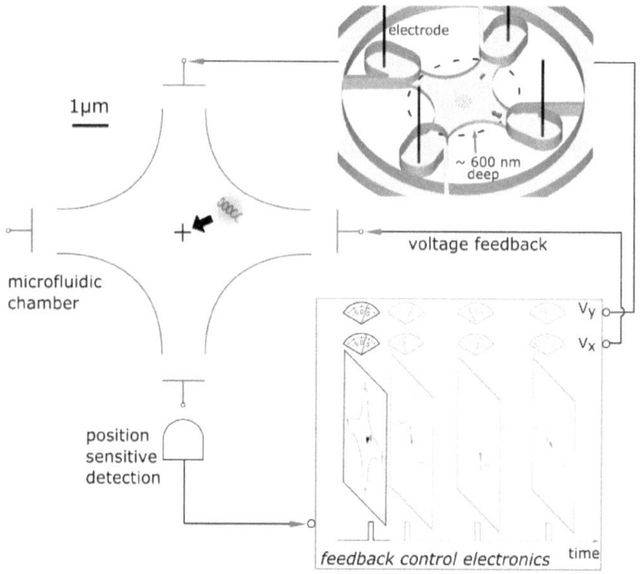

Figure 6.2 Basic principles of the anti-Brownian ELectrokinetic (ABEL) trap. The ABEL trap works as a closed-loop feedback system with three major components: a microfluidic chamber (3D cartoon in blue, 2D projection of the center region on left), a position sensitive detector and a feedback control module. The system continuously tracks the position of a single target molecule (represented here by a piece of double-stranded DNA) via fluorescence emission (red cloud) and calculates a pair of voltages that, when applied to the microfluidic chamber along both Cartesian coordinates, produce electrokinetic forces (black arrow, with the individual components as shaded arrows) that push the molecule towards the center (black "+"). This feedback action is repeated at high speed to maintain trapping of a single target molecule. Modified with permission from Wang, PhD Thesis, Stanford University, 2014.

The prototype of an ABEL trap used four photoresistors taped directly on a TV monitor, in a "magic-wand" arrangement to apply feedback voltage in response to a micron-sized bead's motion (Cohen 2006). The first-generation apparatus (Cohen and Moerner 2005, 2006), which used a fast camera to detect object position, succeeded in trapping 20 nm fluorescent beads and large biomolecules such as λ-DNA, using a commercial inverted microscope, centroid-finding software, and a polydimethylsiloxane (PDMS)-based fluidic chamber. Almost simultaneously, Shapiro's group at the University of Maryland independently developed a system that utilizes video microscopy and feedback electrokinetic control to steer multiple micron-sized objects through arbitrary trajectories (Shapiro et al. 2005). The second-generation ABEL trap utilized a revolving beam position-sensing scheme, similar to Enderlein's earlier proposal (Enderlein 2000), to achieve a low latency feedback. At the same time, a confocal detection geometry and low-background substrates (fused silica) were used to improve the signal-to-background ratio (Cohen and Moerner 2008). These improvements enabled reliable capture of a variety of biomolecules and detailed studies of their biophysical and biochemical properties (Bockenhauer et al. 2011; Goldsmith et al. 2011; Goldsmith and Moerner 2010; Jiang et al. 2011). The third generation of traps, which will be discussed in the remainder of this chapter, utilizes optimized beam-scanning patterns and sophisticated signal processing units to conduct photon-by-photon trapping (Fields and Cohen 2011; Wang and Moerner 2010, 2011). In the most advanced version of the trap, individual fluorescent dye molecules ~1 nm in size, can be captured for several seconds. With the single-dye limit having been reached, any biomolecule (fluorescently labeled), can now be trapped.

It is worth noting that parallel to the evolution of the ABEL trap, many groups developed feedback tracking techniques that use similar ideas. These techniques actively move the sample, usually by a piezo stage, so that the object of interest always stays in the field of view. Conceptually, these approaches are similar to Berg's original apparatus (Berg 1971) for single bacterial cells but were pushed to handle submicron objects. Notable examples include work from Yang Lab (Cang et al. 2006; Hu Cang, Xu, and Yang 2008; Welsher and Yang 2014), the Werner lab (Keller et al. 2014; Lessard, Goodwin, and Werner 2007), the Mabuchi lab (Berglund and Mabuchi 2005; Limouse et al. 2017; McHale and Mabuchi 2009), the Bewersdorf lab (Juette et al. 2013; Juette and Bewersdorf 2010), the Lamb lab (Dupont and Lamb 2011; Katayama et al. 2009), the Davis lab (Germann and Davis 2014), the Yeh lab (Perillo et al. 2015), the Andersson lab (Ashley et al. 2016; Ashley and Andersson 2015), and others. Although the actuation speed is much slower than that of the ABEL trap (~1 ms to move the stage compared to ~1 μs to induce electrokinetic motion), active tracking offers the advantage of the ability to conduct the experiment in cells. Recently, by combining feedback tracking and simultaneous 2-photon imaging (DeVore et al. 2015; Welsher and Yang 2014), it became possible to acquire contextual information surrounding the actively tracked probe. For instance, the Yang group successfully visualized the cellular uptake of a single gold nanoparticle using this approach (Welsher and Yang 2014).

6.2.3 Essential Components of an ABEL Trap

6.2.3.1 Microfluidic sample chamber

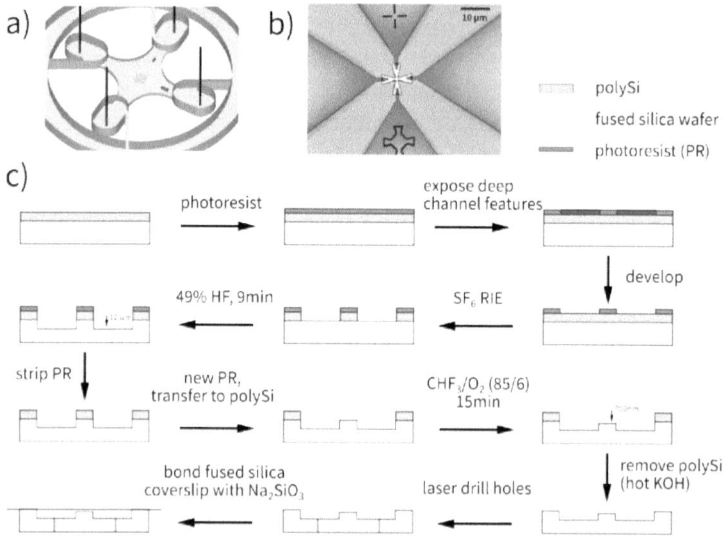

Figure 6.3 Design and fabrication of the microfluidic sample holder for ABEL trap measurements. (a) A 3D model of the chamber. (b) Optical micrograph of the patterned quartz wafer showing the shallow trapping region (yellow), deep channel reservoirs connecting to the shallow region (darker orange) and the bonding regions (light orange). (c) Fabrication process of the fused silica chamber.

The sample chamber (Figure 6.3a) holds the liquid sample while enabling efficient electrokinetic actuation. The basic chamber structure consists of a shallow (~700 nm) region connected with two pairs of orthogonal, deep "reservoirs" (Figure 6.3b). Electrodes are inserted into the reservoirs to induce two-dimensional electrokinetic responses. In this configuration, most of the voltage drop happens across the shallow region, making it possible to induce strong electrokinetic forces with relative low voltages (~10 V).

An outside ring channel connects the deep reservoirs and acts as a pressure relief trench. Both PDMS and fused silica have been used to construct the sample chamber. PDMS is easy to fabricate for prototyping purposes but has a high fluorescence background. On the other hand, fused silica devices have a much lower fluorescence background but are more difficult and costly to make. Key processes required to fabricate the microfluidic sample chambers are illustrated in Figure 6.3c.

6.2.3.2 Position sensing module

A strategy to track the position of a diffusing single molecule in real time is the most critical component of the ABEL trap and has been continuously improved over the years. First generation traps used fast cameras and centroid fitting algorithms to determine position (Cohen and Moerner 2005). This approach worked well for objects of ~100 nm in size but failed to achieve effective trapping of typical proteins (~10 nm). This is mainly due to excessive amount of motion blur (Deschout, Neyts, and Braeckmans 2012; Michalet 2010) that makes localization difficult and inaccurate. To understand this, consider a typical protein molecule with a diffusion coefficient of D = 100 μm^2/s. During a 10 ms exposure (a typical value used in single-molecule imaging), it will be blurred by diffusion with a two-dimensional RMS distance of $(4D\Delta t)^{1/2}$ = 2 μm, which is about five times the size of the (unblurred) point spread function (~400 nm). In other words, keeping the amount of motion blur comparable to the diffraction limit requires an exposure time of ~100 μs (10,000 fps), which is challenging to achieve with a camera. To remedy this, laser-scanning-based methods were subsequently implemented (Berglund and Mabuchi 2004; Cohen and Moerner 2008; Enderlein 2000) to track motion of fast diffusers. In the most advanced version of the scanning method, a focused laser spot undergoes a 32-point "knight's tour" scanning pattern (Figure 6.4a) at the sample plane. This particular beam scanning scheme achieves efficient and fast coverage of a square-like area ~3 μm × 3 μm in size. The scan speed was chosen to be much faster than the diffusion time scale (600 ns per point). When a single molecule is illuminated under this rapidly scanning beam, every photon-stamped beam position provides an estimation of molecule position (Wang and Moerner 2010, 2011)

(Figure 6.4c). Here, position information is encoded on the *timing* of fluorescent photons. By using a point detector with a ~300 ps time resolution instead of an array detector such as a camera, it is possible to sense motion with every detected photon, eliminating motion blur.

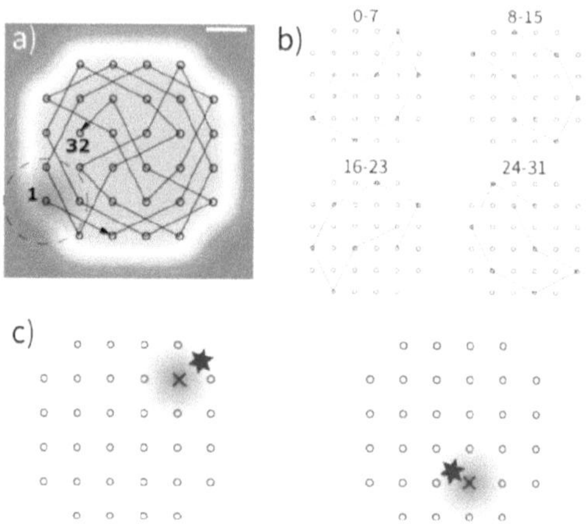

Figure 6.4 Camera-less position sensing by "knight's tour" laser scanning. (a) Trajectory of a 32-point knight's tour scan. Scale bar: 0.5 μm. (b) The full 32-point scan is composed of four 8-point sub frames. Each sub frame covers ~75% of the scan area. (c) The principle of photon-stamped position mapping: at the photon detection instance, the position of the scanning beam (marked by "X") is taken as the estimate of object position (marked by the blue star). Two hypothetical instances are shown. Modified with permission from Wang, PhD Thesis, Stanford University, 2014.

6.2.3.3 Optimal real-time position estimation

Refinement of the photon-stamped position estimates is needed to achieve the ultimate trapping performance. The photon-by-photon position sensing strategy outlined above eliminates motion blur, at the cost of large estimation uncertainty per measurement (Figure 6.5b). If these measurements are used directly for feedback, a large amount of noise will be injected into the feedback loop, making the system unstable. In other words, an inaccurate position estimate used for feedback is equally likely to push the molecule out of the trap. To suppress measurement uncertainty, we utilize established tools from optimal control of stochastic systems with noisy observations, briefly described below.

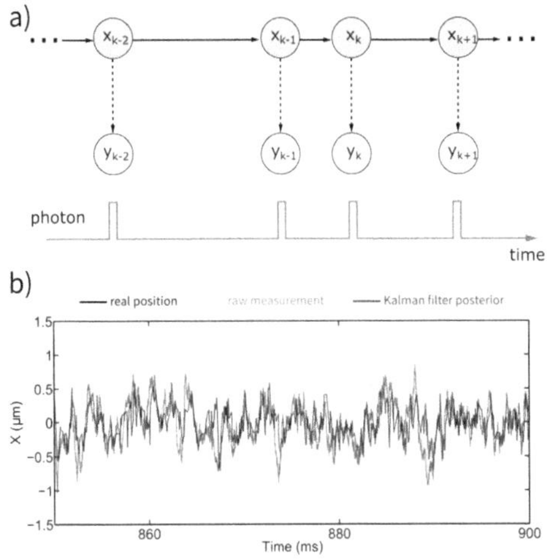

Figure 6.5 Refining position estimation by a Kalman filter. (a) Graphical model representation of photon-by-photon single-molecule tracking. Shaded circles: positions of the target object at instances of measurements. Open circles: raw measurements (photon-stamped beam positions). (b) Simulated trajectory of a single trapped object showing the real position (black), raw measurement (red) and Kalman filtered position estimates (blue). Kalman filter greatly reduces the estimation uncertainty. Modified with permission from Wang, PhD Thesis, Stanford University, 2014.

A graphical representation of the estimation problem is illustrated in Figure 6.5a. Here, x is the real position of the molecule of interest and the quantity we want to estimate from a series of photon-stamped beam positions (y). The time evolution of x follows the Langevin equation,

$$\boldsymbol{x}_{k+1} = \boldsymbol{x}_k + \mu \boldsymbol{U}_k \Delta\tau_k^{fb} + \sqrt{2\Delta t_k D}\,\boldsymbol{Ib} \tag{6.1}$$

where the second term on the right-hand side is the feedback-voltage-induced electrokinetic drift (μ being the electrokinetic mobility) and the third term is the 1D Brownian displacement. The raw, unfiltered measurement y is a function of x, corrupted by measurement uncertainty. In general, we have

$$p(\boldsymbol{y}_k \mid \boldsymbol{x}_k) = f\left(\boldsymbol{x}_k; \Omega\right) \tag{6.2}$$

where Ω summarizes other experimental parameters. The optimal algorithm should utilize all available information to estimate, in real time, the position of the object with minimum uncertainty. We and others (Fields and Cohen 2012) have shown that although an optimal algorithm is computationally expensive, a near-optimal strategy, known as the Kalman filter algorithm, can be easily implemented and achieves an excellent performance in practice.

Under the assumption of high signal-to-background ratio and tight trapping (Wang and Moerner 2010), Eq. (6.2) can be well-approximated by a Gaussian distribution and the optimal estimator that achieves minimum variance can be calculated recursively using a Kalman filter (Welch and Bishop 1995) with the following updating rules:

$$\begin{aligned} \hat{x}_k^- &= \hat{x}_{k-1}^+ + u_{k-1}\mu\Delta\tau_{k\text{-}1} \\ \hat{x}_k^+ &= \hat{x}_k^- + K_k\left(y_k - \hat{x}_k^-\right) \end{aligned} \tag{6.3}$$

where the superscript "-" indicates "prior" estimates, which represent a prediction of molecule position before new measurement information is incorporated and the superscript "+" indicates "posterior" estimates, which represent refined estimates after new measurement information is taken into account. K_k is a weighting factor called the "Kalman gain," to minimize the uncertainty (variance) in the posterior. We need to keep track of and propagate the variances associated with the prior and posterior estimates, which can be done by the following equations

$$\begin{aligned} P_k^- &= P_{k-1}^+ + 2D\left(t_k - t_{k-1}\right) \\ P_k^+ &= \left(1 - K_k\right)^2 \cdot P_k^- + K_k^2 \cdot \left(\frac{w}{2}\right)^2 \end{aligned} \tag{6.4}$$

where w is the beam radius ($1/e^2$ intensity). We can find the Kalman gain that minimizes the posterior estimation variance,

$$K_k^{\text{opt}} = \frac{P_k^-}{P_k^- + \left(\frac{w}{2}\right)^2} \tag{6.5}$$

Because Eqs. (6.3–6.5) are all linear operations, the Kalman filter can be easily implemented on hardware, such as the Field-Programmable-Gate-Array (FPGA).

We consider the Kalman algorithm to be only "sub-optimal" for two reasons. First, optimality requires precise knowledge of the transport coefficients (D and μ), which is not always available. Second, the Kalman filter assumes no background photons, which is not true in experiments. The presence of background photons makes Eq. (6.2) non-

Gaussian, and a generalized filtering scheme is required to compute the minimum variance position estimates. A detailed treatment is given in Supplementary Information of (Wang and Moerner 2014a). However, such an algorithm is much more complex and difficult to implement in real time. Extensive computer simulations also suggest that the gain in trapping performance is minimal when the complex scheme is used.

In summary, combining rapid laser scanning with a hardware-implemented Kalman filter eliminates motion blur while achieving minimum tracking error limited by available information.

6.2.3.4 Signal processing platform

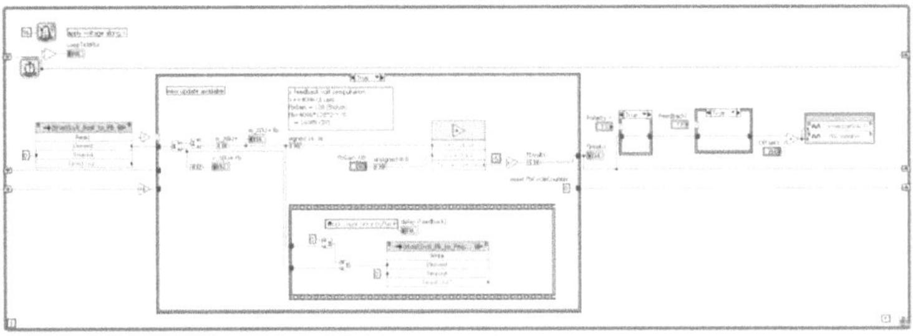

Figure 6.6 A snapshot of the Labview FPGA program to implement the ABEL trap. This specific loop applies feedback voltages to the microfluidic chamber based on the Kalman filter refined position estimates.

The "brain" of the ABEL trap is a piece of hardware that quickly integrates the positional information of the single molecule, determines the direction and magnitude of the feedback kick and applies these voltages. As was discussed previously, a near-optimal feedback strategy is to use the fast beam scanning and a photon-by-photon Kalman filter. We implement the Kalman filter on a National Instrument R Series FPGA board (NI 7842R). The FPGA allows complex state machines to be implemented at high speed and NI's LabVIEW FPGA module provides an intuitive programming interface (Figure 6.6). We achieved a Kalman filter calculation time of about 1 μs on the FPGA.

6.2.3.5 Optical setup

The optical setup of the ABEL trap is similar to a scanning confocal microscope. Figure 6.7 shows the setup of the excitation optics, which creates the rapidly scanning, "knight's tour" pattern shown in Figure 6.4. First, L1 and L2 form a beam expansion/reduction system that controls the size of the laser beam that interacts with the acoustic-optic deflectors (AODxy). Telecentric relays (L3 & L4, L5 & L6, L7 & L8) are utilized to map the pivot plane of AODx to that of AODy and eventually to the focal plane of the objective lens. For detailed design considerations, the reader is referred to the author's PhD thesis (Wang 2014). The detection optics (Figure 6.8) feature a tube lens (f = 75 mm) and a pinhole to limit the detection volume. A telescope system reimages the pinhole plane to the active area of a single-photon counting device (avalanche photodiode).

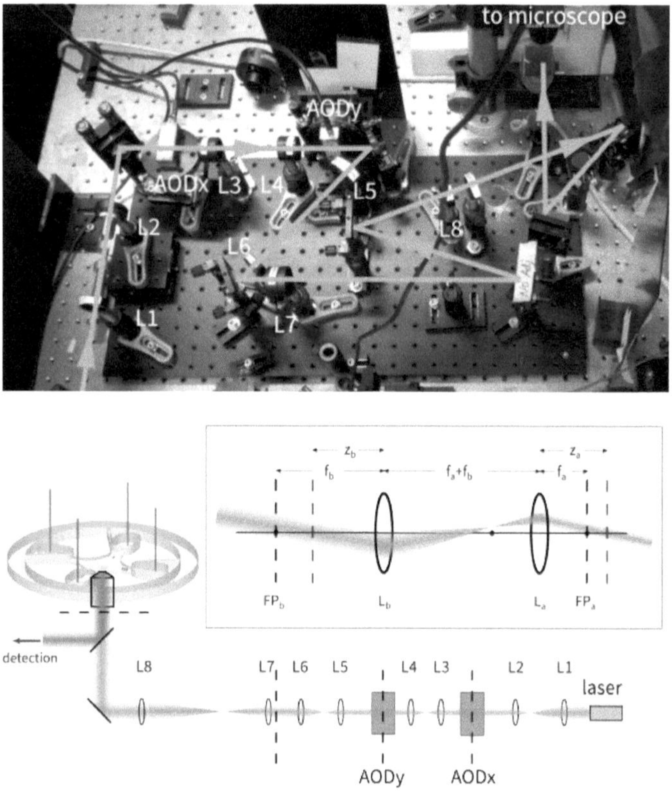

Figure 6.7 Excitation optics. A photo of the setup shows the optical path leading to a commercial inverted microscope base. A schematic is included in the lower panel. Detailed design considerations are discussed in the first author's PhD thesis (Wang 2014).

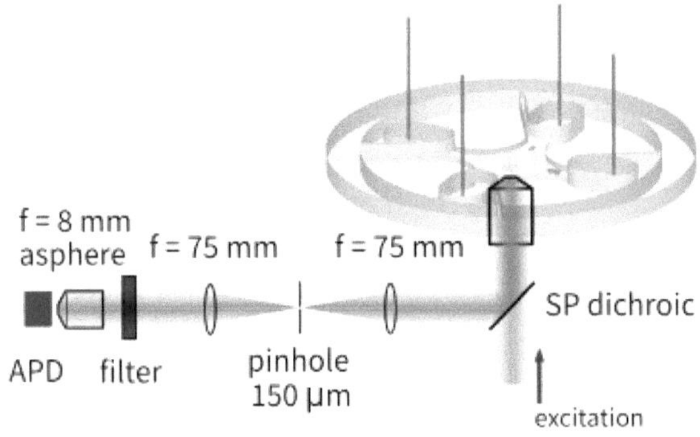

Figure 6.8 Detection optics of the ABEL trap. The pinhole encompasses a circular region of $r{\sim}4$ μm at the sample plane. Modified with permission from Wang, PhD Thesis, Stanford University, 2014.

6.2.3.6 Sample preparation

To perform an ABEL trap experiment, we simply take a solution with ~pM concentration of fluorescent molecules and inject ~0.1 μL of the sample into the microfluidic chamber. After the electrodes are inserted into the chamber, the experiment is ready to start. Single molecules fortuitously diffuse into the scanning area and are captured one after another until the fluorescent tag photobleaches.

Practically, the surface chemistry of the microfluidic chamber needs to be carefully controlled. Two properties of the surfaces strongly influence the experimental outcome. First, the surfaces need to be "non-sticky" to the molecules of interest. Second, the surface charge densities need to be carefully tuned to facilitate electrokinetics. We now discuss these two aspects in more detail. Nonspecific adsorption of biomolecules to the quartz/glass/PDMS surfaces is a critical problem in almost all single-molecule biophysical assays (Fordyce, Valentine, and Block 2008). Most proteins tend to stick to an untreated or potassium hydroxide-activated glass/quartz surfaces in aqueous buffer, presumably due to a combination of electrostatic and hydrophobic interactions. To eliminate this, the interior of the chamber needs to be passivated ahead of the measurement. The most effective method is to coat potassium hydroxide-activated silica surfaces with long chains of polyethylene glycol (PEG). In particular, we find the one-step reaction using 2-[methoxy(polyethyleneoxy) propyl]-trimethoxysilane with a PEG chain length of 20 to 24 units (Gelest, PA) (Sui et al. 2006) particularly convenient and effective. A couple of alternative methods, including blocking by casein, dynamic coating by polyvinylpyrrolidone (PVP) (Milanova et al. 2012) or polyelectrolyte multilayer (Kartalov, Unger, and Quake 2003) have also been tested and found to be effective.

Surface charges of the microfluidic chamber generate electroosmotic flow and should be treated carefully. Surface treatments such as PEGylation and PVP coating greatly suppress electro-osmotic flow, and are desirable in applications where the electrophoretic properties of the analyte are sought after. On the other hand, when dealing with near-neutral molecules, the electrophoretic mobility itself is often too small to induce sufficient feedback actuation. We mitigate this problem by treating the surfaces to support strong electro-osmotic flows. This can be done by sequentially coating the chamber with layers of electrolyte polymers of alternating polarity (polyelectrolyte multilayer) (Decher 1997; Kartalov, Unger, and Quake 2003). In this method, both the terminating polarity and the density of surface charges can be controlled to suit particular applications.

Buffer conditions need to be optimized for ABEL trap experiments. Fluorescent labels often exhibit "blinking," where they switch randomly between "on" and "off" states. When the fluorescent label is "off," no information regarding the molecule's position can be obtained and the feedback mechanism fails, leading to molecule escape. It is thus critical to minimize blinking for prolonged single-molecule capture. Many protocols have been published, for an extensive review, see ref. (Zheng et al. 2014). In our hands, the best trapping performance can be achieved by either one of the following red dyes: Cy5, Alexa 647, Atto633 or Atto647N, with Trolox and oxygen removed (Vogelsang et al. 2008). Caution must be taken when choosing oxygen scavenger systems as commercially available products might contain unwanted DNA processing activities (Senavirathne et al. 2015; Swoboda et al. 2012). In cases where oxygen removal is undesirable, the oxazine dyes Atto655 and Atto680 provide reasonably good performance.

6.3 Selected Applications of the ABEL Trap

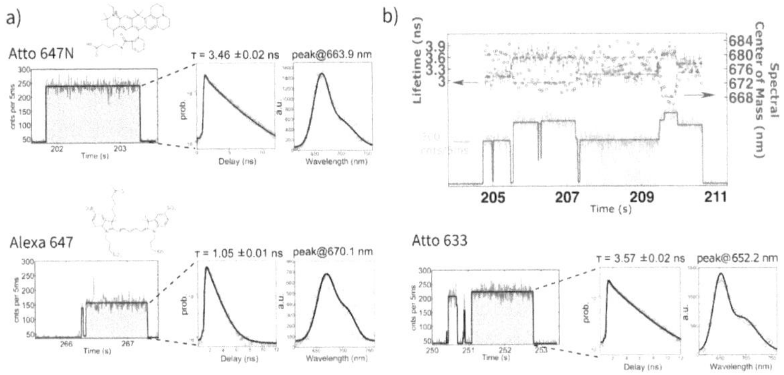

Figure 6.9 Multiparameter spectroscopy on individual fluorophores held in the ABEL trap. (a) Representative traces of trapped single fluorophores with simultaneously recorded excited-state lifetime and emission spectrum. The same excitation power was used in all three cases so the differences in measured intensities reflect the relative brightness with 594 nm excitation. (b) Photodynamics of Atto647N resolved in the ABEL trap. Individual dyes sometimes switch between different emissive states of distinct brightness, excited-state lifetime and emission spectrum. Reprinted with permission from (Wang and Moerner 2013) Copyright (2013) American Chemical Society.

The ABEL trap allows single nanometer-sized objects, including biomolecules and nanoparticles, to be observed in an aqueous solution for extended periods of time, during which an immense amount of quantitative information can be gathered. The platform has been adopted by many labs to study a diverse array of problems across chemistry, physics and biology and new applications are still emerging. Early work on monitoring the dynamics of biomolecules by the Moerner lab is summarized in the account by Wang et al. (Wang et al. 2012). More recent work using the ABEL trap to study photosynthetic antenna proteins yielded new insight on the structure-function relations of this important class of protein complexes (Goldsmith and Moerner 2010; Schlau-Cohen et al. 2013, 2014, 2015; Squires et al. 2019; Squires and Moerner 2017; Wang and Moerner 2015). The Goldsmith lab used single-molecule anisotropy information in the ABEL trap to probe the conformation of intrinsically disordered tau protein in solution (Foote et al. 2019; Manger et al. 2017). The

Börsch lab used the trap (M. Dienerowitz, Dienerowitz, and Börsch 2018) to monitor the rotation of the FoF1 ATP synthase (Dienerowitz et al. 2016; Su et al. 2015). Parallel to those investigations which are of a biochemical nature, the ABEL trap has also proven to be a valuable tool for physical sciences, with applications ranging from statistical mechanics (Cohen 2005; Jun, Gavrilov, and Bechhoefer 2014), polymer physics (Cohen and Moerner 2007) to nitrogen vacancy centers (Kayci, Chang, and Radenovic 2014; Kayci and Radenovic 2015). Below we highlight a few case studies.

6.3.1 Reaching Ultimate Limit: Trapping Single Organic Fluorophores in Solution

Individual copies of fluorophores represent the ultimate limit of any trapping modality. They are about ~1,000 Da in molecular weight and ~1 nm in size. So far, only the ABEL trap has reached this limit. Figure 6.9a shows representative trapping traces of three widely used fluorescent dyes: Atto633, Atto647N, and Alexa647. To stably trap these single fluorophores, both blinking and photobleaching must be suppressed (Vogelsang et al. 2008). Specifically in these experiments, oxygen was removed by a protocatechuate system (Aitken, Marshall, and Puglisi 2008) and Trolox (Rasnik, McKinney, and Ha 2006) at ~3 mM concentration was used to create a reducing and oxidizing environment (Cordes, Vogelsang, and Tinnefeld 2009). Without these buffer components, very few molecules lasted more than 1 s, presumably due to dark-state formation. In the examples given in Figure 6.9a, the relative intensities reflect the brightness differences of the dyes under the same illumination conditions, 594 nm excitation.

Also shown in Figure 6.9a are simultaneously recorded excited lifetimes and emission spectra of trapped single molecules. Typically, we were able to harvest >100,000 photons from these molecules for spectroscopic characterization. This represents three orders of magnitude improvement over free-diffusion-based methods (Widengren et al. 2006). Taking advantage of this unique capability of the ABEL trap, we determined the excited-state lifetimes of those trapped molecules with <50 ps precision and resolved the full emission spectrum of single molecules in solution. The measured lifetimes and spectral peak positions, when averaged over all measured molecules, agree well with manufacturer's values for these dyes. However, the authors saw significant molecule-to-molecule heterogeneity in Atto633 and Atto647N (Wang and Moerner 2013, 2014b). Even detailed features in the spectrum, such as the vibronic progression, can be well-resolved. Moreover, the seconds-long observation window enables direct visualization of slow photodynamics in solution.

Surprisingly, our measurements revealed that some Atto633 and Atto647N molecules occasionally switch emission states during their residences in the trap. A representative Atto647N molecule that displays this switching behavior is shown in Figure 6.9b. To better characterize the different emissive states observed, we conducted multiparameter mapping of the emissive states using data from ~1,000 molecules. This analysis revealed that Atto647N switches between three major states with distinct brightness, excited-state lifetime, and spectra (Wang and Moerner 2013). Atto633 exhibits a similar three-state switching behavior and also infrequently populates a number of rare, but distinctive substates (Wang and Moerner 2014b). Some of these transitions were found to be light-driven while others seemed to be spontaneous. The detailed mechanism of the observed photodynamics remains to be investigated. In contrast, another widely used dye, Alexa 647, was observed to display only one homogeneous emission state.

The capability to trap single organic fluorophores marked the maturation of the ABEL trap technology. Any biomolecule can now be studied in the trap when coupled to a good fluorescent label. The revelation that some dyes exhibit multiple emission states in solution highlights the importance of understanding dye photophysics when using dyes as reporters for nanoscale dynamics (Taekjip Ha and Tinnefeld 2012).

6.3.2 Dissecting Pigment Organization of Single Biliproteins in Solution

The ABEL trap with multiparameter fluorescence recording capability proved to be a powerful tool to illuminate the optical properties of photosynthetic antenna protein complexes in solution (Schlau-Cohen et al. 2014). These photosynthetic antenna proteins are evolutionarily tuned pigment–protein complexes that absorb specific bands of the solar radiation and transfer the absorbed energy across long distances (~10 nm) to fuel the reaction centers (Croce and van Amerongen 2014). Interestingly, nature uses relatively similar pigment building blocks to achieve diverse functions, and the organizational principles behind this process are not fully understood. Single-molecule spectroscopy has become a powerful method to investigate these systems (Kondo, Chen, and Schlau-Cohen 2017) but traditional techniques introduce significant perturbation to the delicate structural-functional balance of these proteins and the results are often complicated by experimentally introduced artifacts. The ABEL trap conducts measurements on single copies of the proteins directly in solution and is thus less perturbative.

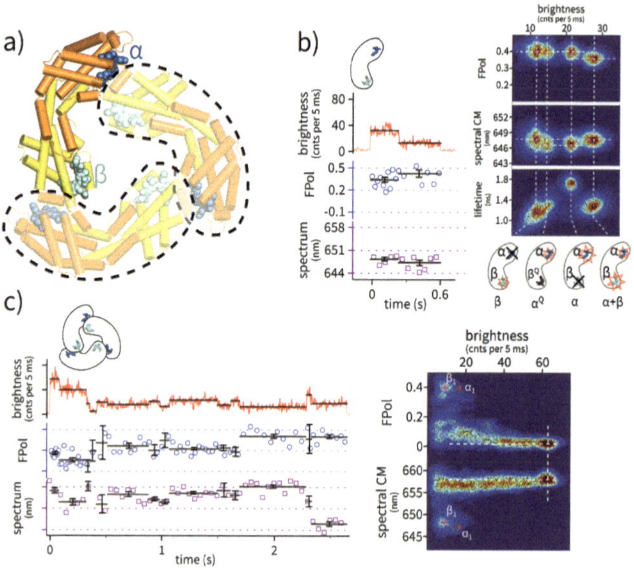

Figure 6.10 Dissecting the pigment organizational principles of allophycocyanin (APC) in the ABEL trap. (a) Crystal structure of APC (PDB: 1ALL) showing the geometric arrangement of the pigments. (b) Multiparameter fluorescence dynamics of APC monomers. Left: an example molecule. Right: resolved emission states from ~2,000 molecules. State identifications are illustrated at the bottom. (c) Multiparameter fluorescence dynamics of APC trimers. Left: an example molecule. Right: resolved emission states from ~800 molecules. Pigment properties of the monomer are overlaid on the plot with "+" symbols. Reprinted with permission from (Wang and Moerner 2015).

Successful measurements have been conducted on LH2 (purple bacteria) (Schlau-Cohen et al. 2013), PCP (dinoflagellates)

(Bockenhauer and Moerner 2013), LHCII (higher plant) (Schlau-Cohen et al. 2015) and phycobiliproteins (cyanobacteria) (Squires and Moerner 2017; Wang and Moerner 2015). Here, we describe one particular set of measurements (Figure 6.10) to dissect the optical properties of individual pigments on an important phycobiliprotein: allophycocyanin (APC) (Wang and Moerner 2015).

In cyanobacteria, APC plays an important role in light-harvesting, energy transfer, and photoprotection. Each monomer of APC covalently binds two (referred to as α and β) phycocyanobilin pigments in similar but distinct protein environments. Monomers self-assemble to form trimers (Figure 6.10a). During the self-assembly process, the absorption spectrum of the protein redshifts is ~30 nm. Due to the functional importance of this red shift, we set out to characterize the optical properties of individual pigments on the protein in both the monomer and trimer using the ABEL trap. To maximize the amount of spectroscopic information, we measured multiple fluorescence parameters including brightness, excited-state lifetime, fluorescence polarization, emission spectrum, and photon antibunching. Some of these parameters were measured simultaneously.

Monomers were observed to show digital transitions between distinct brightness levels with small changes in polarization and emission spectrum (Figure 6.10b). Detailed analysis of the emission levels revealed four states in the multidimensional parameter space. Three of these states were expected: the pristine monomer (i.e., both α and β not bleached), α (β bleached), and β (α bleached) and our multiparameter measurement allowed spectroscopic characterization of these states with unprecedented precision. We found that α and β have slightly different emission signatures. Moreover, we discovered a new (unexpected) state, which we assigned to be α quenched by the photoproduct of β, based on excited-state lifetime and emission spectrum signatures. Although evidences of quenching sites were presented in early single-molecule studies (Ying and Xie 1998), we were able to pinpoint the site location of the quencher.

Similar measurements on trimers of APC resulted in a complicated multiparameter map (Figure 6.10c). However, most of the data can be explained by a simple model that featured a pronounced spectral redshift of the α pigment upon trimer formation and the occasional generation of localized quenching site on the β pigment upon photoexcitation. Taken together, we were able, for the first time, to measure the emission properties of individual pigment sites on the protein and understand how the protein reorganizes to generate synergistic properties upon self-association. These insights are not only valuable for understanding the photosynthetic machinery of cyanobacteria but also provide guidance to the intelligent design of artificial energy harvesting devices.

6.3.3 Sensing Biomolecular Interactions by Single-Molecule Transport

In the previous two examples, we have demonstrated how the ABEL trap enables synchronous dynamics of multiple fluorescence parameters to be recorded on single molecules in solution. These measurements reveal the internal states of a nano-emitter and the interconversions between these states on a ~1 s timescale, with unprecedented versatility and precision. So far, we have restricted ABEL trap studies to intrinsically fluorescent molecules. Most biomolecules that make life happen, such as proteins and nucleic acids, are not directly observable but can be detected indirectly by attachment of fluorescent moieties. When studying these molecules, it is not the fluorescent properties of the probe that is of interest but the intrinsic behaviors of the biomolecules.

The general approach to study the intrinsic behaviors of biomolecules using fluorescence is to engineer the probe so that its emission properties are modulated by the biomolecule's behaviors (e.g., FRET senses conformational change by distance-dependent fluorescence quenching). On the other hand, to broaden the utility of single-molecule measurements, it is much more desirable to sense directly the intrinsic physical properties (e.g., size, shape, charge, etc.) of the single biomolecules themselves. Here, we demonstrate how the ABEL trap can estimate parameters related to size and charge of single biomolecules in solution (Wang and Moerner 2014a). We believe that this critical advance opens up new directions in single-molecule sensing.

To sense size and charge, we focus on how a single molecule moves. When held in the ABEL trap, a single molecule's residual motion is governed by a combination of diffusion and electrokinetic drift. We developed statistical analysis tools that extract diffusion coefficient (D) and electrokinetic mobility (μ) from observed photon-by-photon single-molecule trajectories (Figure 6.11a). This new analysis module adds two powerful extensions of the ABEL trap platform: (1) the ability to measure diffusivity and mobility of a single molecule with unprecedented precision and (2) the ability to measure time-dependent dynamics in diffusion coefficient and mobility.

We demonstrated the power of these new single-molecule observables using well-studied biomolecules. First, we monitored the dissociation pathway of APC, the subject of the aforementioned spectroscopic studies (Figure 6.11b). Here, we estimated the molecular diffusivity and mobility of individual trapped proteins and plotted them on a D–μ parameter space. In the case of the crosslinked control, where the protein is not capable of dissociation, we resolved one population representing the trimer. When we measured wild-type proteins, we observed two populations that represent the trimer and monomer species and differ mainly in diffusivity. The ability to differentiate the different oligomerization states of a protein by their hydrodynamic sizes will fill a critical gap in dissecting and quantifying protein assembly pathways in the small-oligomer regime.

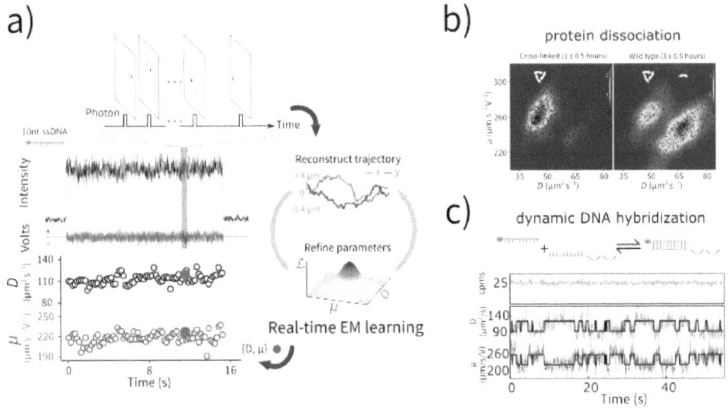

Figure 6.11 Extracting size and charge sensitive motion parameters of trapped single molecules. (a) Principles of the algorithm. A time slice of trapping data is subjected to a machine learning algorithm to extract the diffusion coefficient (D) and electrokinetic mobility (μ). (b) Resolving a monomer-trimer mixture along a protein dissociation pathway by mapping of single-molecule motion parameters (D and μ). (c) Direct visualization of the size and charge fluctuations induced by DNA hybridization and melting events. Binding of a complementary strand lowers D and increases μ while melting increases D and decreases μ. Reprinted with permission from Ref. (Wang and Moerner 2014a) Copyright Nature Publishing Group.

We next used the method to sense bimolecular interactions. In this experiment, we trapped a dye-labeled single-stranded DNA and monitored its diffusion coefficient and mobility in the presence of its unlabeled complementary strand. Two-state, anticorrelated fluctuations of D and μ were observed (Figure 6.11c) that visualized the dynamic binding and unbinding of the complementary strand. This experiment opens the door to direct observation of protein–DNA and protein–protein interactions at the single-molecule level.

6.4 Summary and Perspective

The ABEL trap is now a mature technology to examine dynamics in solution at the nanoscale. Compared to existing technologies, the ABEL trap offers several unique advantages. First, it allows single molecules to be measured directly in biologically compatible buffer solutions, without attaching them to any surfaces. In this mode of probing, biomolecules retain complete rotational and diffusional degrees of freedom. Second, it is fully compatible with existing fluorescence sensing modalities (i.e., FRET (Roy, Hohng, and Ha 2008), PIFE (Hwang and Myong 2014)) while providing additional information on molecular size and charge. Here, we outline future technical development of the ABEL trap, as well as its applications in biophysical research. The application of the ABEL trap to study statistical physics is reviewed in (Gavrilov and Bechhoefer 2017).

New capabilities: Current operation of the trap lacks the means to introduce new reagents or change the chemical and physical environment around the molecule. We envision the next-generation trapping device will feature automated sample delivery and mixing capabilities (Wunderlich et al. 2013) and rapid temperature control (Holmstrom and Nesbitt 2010). These technical advances will open the door to a much wider class of scientific problems including protein folding and unfolding, T-jump, and other nonequilibrium measurements. The newly developed MINFLUX (Balzarotti et al. 2017) imaging modality promises higher localization precision for a given number of photons detected and could further improve the feedback efficiency.

While the ABEL trap technology has existed for more than a decade, using the technology to study biomolecules is only now starting to show great promise and is expected to yield rich scientific insight in the near future. We point out several future directions in which the ABEL trap will be particularly useful and complementary to existing approaches.

Protein–protein interactions: The capability to measure diffusion coefficients of single molecules over long periods of time allows one to detect binding and unbinding events directly. Another attractive aspect is that only one molecular species needs to be fluorescently labeled and kept at single-molecule concentration, other species, generally those that interact with the labeled target, do not need to be labeled and thus can be present at arbitrarily high concentrations. It is thus possible to monitor multiprotein interactions in vitro one molecule at a time. Another important application is to track protein oligomerization and aggregation pathways, which are fundamentally important processes in protein function and disease pathogenesis (e.g., Alzheimer's, cataracts, etc.) but are poorly understood due to lack of sensitive methods in the small-oligomer regime. The ABEL trap provides a direct route to quantify the oligomer distribution by measuring the size of each molecule one at a time.

DNA–protein interactions: The interactions of DNA and proteins play key roles in essential life processes such as transcription, replication, and genome maintenance. The ABEL trap offers unique capabilities to dissect these nanoscale transactions at the individual molecule level. Together with high-resolution fluorescence imaging techniques (Cohen and Moerner 2007), it is possible to visualize how large DNA loops are extruded by proteins (Goloborodko, Marko, and Mirny 2016). Combining the ABEL trap with single-molecule FRET measurements will simultaneously yield multidimensional information on molecule size, charge, and intramolecular distances. These new measurements are expected to provide a rich mechanistic insight of fundamental life processes.

Sensing of single-molecule charge: The amount of charge on a protein is a fundamental property in biophysical science. Although the charge state of a protein has long been recognized to play important roles in protein biochemistry (Gitlin, Carbeck, and Whitesides 2006), it has attracted little experimental attention at the single-molecule level, largely due to lack of suitable techniques with which to measure molecular charge. The ABEL trap allows the charge state of a single biomolecule to be directly measured, by analyzing its motion under electric field. We envision that it will soon become possible to monitor a single molecule's charge together with other degrees of freedom. Monitoring biochemical processes that change the isoelectric point of a protein, such as phosphorylation, will become possible.

New discoveries are often directly enabled by new technologies. We are confident that in the near future, new scientific insights will be gained by the unique capabilities of the ABEL trap.

Appendix: A Tutorial Protocol: Trapping Single Molecules of 10-Nucleotide Single-Strand DNA in Solution

Purpose

The following protocol describes the procedures with which to conduct a typical ABEL trap experiment. It is demonstrated with 10-nucleotide single-stranded DNA labeled with Atto647N dye at the 5' end as a standard analyte. Upon completion of this protocol, the researcher can adapt the procedures to trap other biomolecules of interest (Wang et al. 2012). To implement this protocol with other biomolecules, one should consider following two requirements. First, the biomolecule needs to be fluorescently labeled (or intrinsically fluorescent) and detectable at the single-molecule level. Second, surface coating conditions need to be optimized specifically to the molecule of interest.

Materials

- **Wet Lab Equipment**
  Laminar flow hood (Airscience HLF-XT-24)
  Pure nitrogen stream source (Airgas NI HP300)
  Observation Microscope (Amscope T690B-PL)
  Bonded ABEL trap microfluidic chip (custom made, see Figures 6.3 and 6.12)
  Sonicator (Crest CP230D)
- **Consumables**
  Metal tweezers (Dumostar 2A)
  Glass beaker (Pyrex)
  Glass pipette (Fisher)

0.2 μm SFCA filter (Fisher)
Lens tissues (Tiffen)
Ethanol (Sigma-Aldrich, HPLC grade)
Acetone (Sigma-Aldrich, HPLC grade)

- **Reagents**
  Nanopure water (18.2 MΩ cm^{-1})
  Sulfuric acid (Sigma-Aldrich)
  Hydrogen peroxide (30% wt, Sigma-Aldrich)
  1M Potassium hydroxide (Sigma-Aldrich)
  2% Alconox solution
- **Sample**
  Fluorescently labeled ssDNA (5'-Atto647N-AACTTGACCC-3', supplied by IDT), dissolved at a 45 μM concentration in the storage buffer.
  Storage buffer: 20 mM HEPES pH 8, 100 mM NaCl
  Trapping buffer: 20 mM HEPES pH 8, 100 mM NaCl, 2 mM Trolox
  Oxygen Scavenger system: protocatechuate-3,4-dioxygenase (OYC America) Protocatechuic acid (Sigma Aldrich)
- ABEL Trap Hardware & Software

Safety and precautions

- Piranha cleaning (3:1 sulfuric acid:hydrogen peroxide) should be carried out inside a fume hood with appropriate PPE (lab coat, goggles, double layers of gloves). Piranha solutions are extremely corrosive, reactive, and potentially explosive. Please refer to Standard Operation Procedures (SOP) in your own institute for detailed safety precautions. Rigorous training is usually required to carry out this procedure.
- 1 M Potassium hydroxide solution is corrosive and must be handled with caution.

Composition of the ABEL Trap Chip

A specially designed and fabricated microfluidic chip is required to conduct ABEL trap experiments. Please refer to the main text and Ref. (Cohen and Moerner 2008) for details. Briefly, the trap is composed of a fused silica coverslip at the bottom and a fabricated piece (also made from fused silica) attached on top. The top fabricated piece contains four through-holes which allow sample injection and electrode insertion. See Figure 6.12 for the schematic of a trap chip.

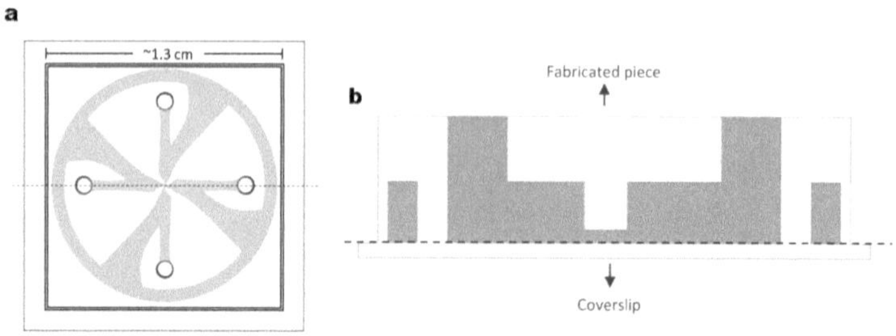

Figure 6.12 Schematic of a microfluidic chip designed for ABEL-trap experiment. (a) Top view of the chip. Outer square is the fused silica coverslip. The inner square represents the fabricated fused silica piece attached on top of the coverslip. Four small circles are the holes on the fabricated piece which are used as ports to fill samples inside the chamber. The gray region marks the channel between the coverslip and the fabricated piece that is filled with sample. White areas represent regions where the coverslip and fabricated piece are bonded. (b) Cross-sectional view of the chip along the dashed line shown in panel a. Bottom thin rectangle is the coverslip and the top thick rectangle is the fabricated piece. The PDMS reservoir is not shown.

Methods

A. Cleaning and Other Preparation of the ABEL Trap Microfluidic Chip

Surface chemistry treatment is extremely critical and needs to be strictly performed ahead of the measurement. Thus, excessive and rigorous cleaning of the microfluidic chip and the reservoir prior to surface preparation and sample injection is critical.

1. **Preparing the microfluidic chip for cleaning**

   A trapping chip is stored in Nanopure water prior to use. It should be handled from the fabricated piece using metal tweezers and blown dry using a nitrogen stream. The trap is then placed onto a lens tissue that is taped to the working surface inside a laminar flow hood. The nitrogen stream is directed into the ports of the trap to dry remaining water inside the trap. The trap is then visually checked under a light microscope. Repeat drying with nitrogen until no wet areas are observed anymore.

2. **Cleaning the chip using Piranha**

   To completely remove fluorescent and other impurities, the chip should be effectively cleaned using a Piranha solution. The following section gives a general description of the steps involved in Piranha cleaning and does not replace the SOP of the Piranha procedure. Proper training is required and should be administrated by your institution when handling Piranha cleaning.

   After ensuring that the trap is completely dry, it is then transferred to a fume hood. A beaker of Piranha solution is prepared by adding 10 mL hydrogen peroxide to 30 mL sulfuric acid. One drop of the freshly prepared Piranha solution is immediately dropped onto one port of the trap. Capillary action will draw the solution inside the chip so that the chip is filled. Once the inner chamber is completely filled with Piranha, the whole chip is submerged in the beaker with Piranha solution. The whole chip is incubated for 40 minutes and is then transferred to the storage beaker filled with water and later rinsed using nanopure water and completely dried.

3. **Cleaning the chip with 1 M potassium hydroxide**

   Chip incubation with 1 M potassium hydroxide (KOH) solution renders the interior of the chip hydrophilic, which is desired for downstream surface coating procedures and sample loading. To carry out this step, 10 mL KOH (1M concentration) is filtered using a 0.2 μm filter into a clean beaker. The interior of the chip is filled with 10 μL of filtered KOH. The filled chip is submerged in the KOH beaker and incubated for 10 minutes. The chip is then removed from the KOH beaker, rinsed extensively with water and dried. If the sample will not be loaded within 5 minutes, keep the chip filled with water.

B. Sample Preparation

We used a strand of 10 nucleotide single-stranded DNA labeled with Atto647N dye at the 5′ end (5′-Atto647N-AACTTGACCC-3′, prepared by IDT) as a model analyte. Initial concentration of this DNA stock is 45 μM dissolved in storage buffer (20 mM HEPES pH 8, 100 mM NaCl). The DNA sample is diluted to a 5 pM final concentration in a trapping buffer (20 mM HEPES, 2 mM Trolox, 100 mM NaCl) that contains an oxygen scavenger system (50 nM Protocatechuate 3,4-dioxygenase and 2 mM protocatechuic acid). About 0.5 μL of the final solution is needed to fill the chip. Finally, mount the chip on a sample holder and secure in place with clamps.

C. Setting Up the Trap

Transfer the sample holder to the microscopy setup and place the sample holder on the microscope stage. Run the custom written LabVIEW program (Figure 6.14). Turn on the trapping apparatus (e.g., Laser, piezo stage, etc.) and peripheral electronics (e.g., eyepiece camera, etc.). Use the position controls on the stage to align the sample at the center of the laser illumination spot, as shown in Figure 6.13. Insert electrodes into the ports on the PDMS reservoir according to the marked polarities. Optimize the focus and enable the focus lock.

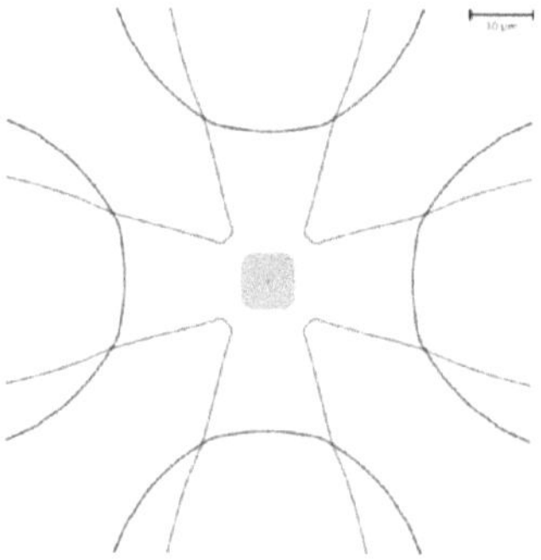

Figure 6.13 Schematic showing proper alignment of the trapping chamber to the laser illumination area. Green square represents the laser illumination area. The cross-like area around the laser spot is the shallow region of the chamber (~0.7 μm depth) where trapping takes place. Areas surrounded by the half circles near the edges represent the deep regions (~10 μm) of the chamber.

D. Performing the Measurements

After the sample is properly positioned, single-molecule trapping experiments may begin. First, the room lights have to be turned off before the single-photon detector is turned on. Adjust the laser excitation power to a suitable level (~50 μW before the microscope objective). Switch the detection path from the eyepiece/camera to the single-photon detector but leave the feedback off. At this point, individual Atto647N-ssDNA molecules randomly diffuse in and out of the laser illumination area and should give rise to transient spikes of fluorescent signals similar to that demonstrated in Figure 6.15a. If no such behavior is observed, troubleshooting becomes necessary. Possible causes may include (a) misalignment of the detection optics (b) ineffective surface preparation that led to nonspecific adsorption of the target molecules by the surface of the trapping chamber (c) the chamber is blocked and prevents molecules entering the shallow trapping region. Extensive trouble shooting procedures are outside the scope of this tutorial protocol.

After confirming the presence of individual freely diffusing molecules, feedback trapping can be turned on. To do so, turn on the high-voltage amplifiers and use the software interface (Figure 6.14) to enable feedback.

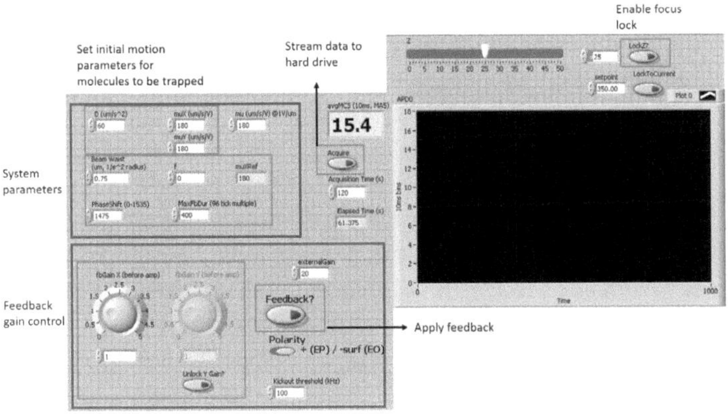

Figure 6.14 Labview control interface of ABEL trap operation.

Figure 6.15b shows a typical dataset: single molecules of 10-nucleotide Atto647N-ssDNA are trapped one after another for tens of seconds, limited by the photobleaching time of the dye Atto647N. Comparing panel 'b' to panel 'a' in Figure 6.15, it is apparent that the ABEL trap extends the observation time of single molecules in solution by a significant amount.

E. Finishing Up

After completion of the measurement, control software and peripheral electronics are switched off, electrodes are taken out, detached and put back into a storage flask filled with pure water. The objective is lowered and cleaned using lens tissue and ethanol. The microfluidic chamber is taken into a laminar flow hood, rinsed and disassembled for future use. In particular, solution from each port of the reservoir is removed using pipette. The reservoir is then peeled off by pressing down on the fabricated piece of the chip using metal tweezers and put away in its storage baker. The trap is blown dry using a nitrogen stream and rinsed with water repeatedly until water completely goes in and it is submerged into water in its storage beaker. The same trapping chamber can be reused repeatedly.

F. **Data Analysis**

A wealth of information can be extracted from trapped single molecules in solution. Here, we briefly describe the data processing procedure and give examples of common analysis.

The first step in the analysis is to extract events corresponding to individual molecules from raw data traces (Figure 6.15b), by identifying the start and end times of every molecule. Toward this end, we use a change point detection algorithm (Watkins and Yang 2005) to detect automatically the abrupt intensity jumps when a molecule enters the trap (from low to high, Figure 6.15b) and when a molecule escapes (from high to low, Figure 6.15b). Once the trace is segmented, data associated with single molecules, which include the intensity as a function of time and the feedback voltages required to keep the molecule trapped, are saved as individual files for further processing.

Figure 6.15c–e illustrates several diagnostic analyses. Panel c is a histogram of trapping times. This is the total duration a single molecule is observed in the trap. In this case, many ssDNA molecules are trapped for tens of seconds. Panel d is a histogram of single-molecule brightness, calculated by averaging the fluorescence intensity (counts per bin) over all time bins. There is significant heterogeneity in this data set (i.e., several brightness peaks), presumably due to photodynamics of the dye Atto647N. Similar heterogeneity was also observed at the single dye level (Figure 6.9). For every trapped single molecule, the time-dependent feedback voltages contain information of the molecule's transport coefficients. Panel e plots the extracted diffusion coefficient (D) and electrokinetic mobility (μ) of individual molecules in a two dimensional parameter space, together with the estimated density distribution. Here, one population with D ~150 $\mu m^2/s$ and μ ~ 95 μm/s/V is observed, which corresponds to the ssDNA molecules measuring 10 nucleotides in length. The diffusion coefficient and the mobility reveal size and charge information of the analyte and can serve as new variables for single-molecule measurements (Figure 6.11).

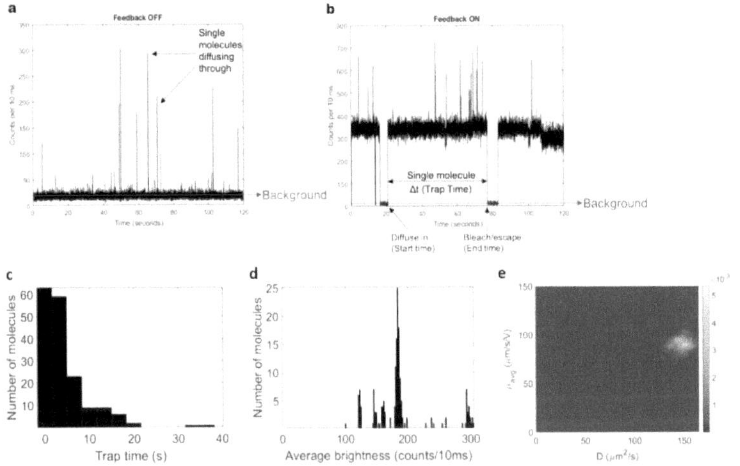

Figure 6.15 Example traces and analysis. (a) Representative brightness trace (photon counts per 10 ms) when no feedback is applied. Single spikes represent freely diffusing single molecules in and out of the laser excitation region. (b) Representative brightness trace with feedback applied. Stable brightness levels that last tens of seconds represent a single molecule (10 nucleotide ssDNA) trapped at the center of the chamber as a result of feedback actuation. (c) A histogram of trapping time, as defined in panel b. (d) A histogram of single-molecule brightness. (e) 2D distribution of diffusion coefficient (D) and mobility (μ) of all trapped molecules. Black dots represent individual molecules, color map represents estimated density.

References

1. Aitken CE, Marshall RA, Puglisi JD (2008). An oxygen scavenging system for improvement of dye stability in single-molecule fluorescence experiments, *Biophys J,* 94(5), 1826–1835.
2. Ashkin A, Dziedzic JM, Bjorkholm JE, Steven C (1986). Observation of a single-beam gradient force optical trap for dielectric particles, *Opt Lett,* 11(5), 288–290.
3. Ashley TT, Andersson SB (2015). Method for simultaneous localization and parameter estimation in particle tracking experiments, *Phys Rev E,* 92(5), 052707. https://link.aps.org/doi/10.1103/PhysRevE.92.052707 (May 2, 2017).
4. Ashley, TT, Gan EL, Pan J, Andersson SB (2016). Tracking single fluorescent particles in three dimensions via extremum seeking, *Biomed Optics Express,* 7(9), 3355. https://www.osapublishing.org/abstract.cfm?URI=boe-7-9-3355 (May 2, 2017).

5. Balzarotti F, et al. (2017). Nanometer resolution imaging and tracking of fluorescent molecules with minimal photon fluxes, *Science,* 355(6325), 606–612. http://science.sciencemag.org/content/sci/355/6325/606.full.pdf (November 15, 2016).
6. Berg HC (1971). How to track bacteria, *Rev Sci Instrum,* 42(6), 868–871. http://aip.scitation.org/doi/10.1063/1.1685246 (April 24, 2017).
7. Berglund A, Mabuchi H (2005). Tracking-FCS: fluorescence correlation spectroscopy of individual particles, *Opt Express,* 13(20), 8069–8082.
8. Berglund AJ, Mabuchi H (2004). Feedback controller design for tracking a single fluorescent molecule, *Appl Phys B,* 78(5), 653–659.
9. Berthelot J, et al. (2014). Three-dimensional manipulation with scanning near-field optical nanotweezers, *Nat Nanotechnol,* 9(4), 295–299. http://www.nature.com/doifinder/10.1038/nnano.2014.24.
10. Betzig E, et al. (2006). Imaging intracellular fluorescent proteins at nanometer resolution, *Science,* 313(5793), 1642–1645.
11. Bockenhauer S, et al. (2011). Conformational dynamics of single G protein-coupled receptors in solution, *J Phys Chem B,* 115(45), 13328–13338.
12. Bockenhauer SD, Moerner WE (2013). Photo-induced conformational flexibility in single solution-phase peridinin-chlorophyll-proteins, *J Phys Chem A,* 117(35), 8399–8406.
13. Boehm EM, et al. (2016). Quantifying the assembly of multicomponent molecular machines by single-molecule total internal reflection fluorescence microscopy, In *Methods in Enzymology,* http://linkinghub.elsevier.com/retrieve/pii/S007668791630266X (October 15, 2016).
14. Boukobza E, Sonnenfeld A, Haran G (2001). Immobilization in surface-tethered lipid vesicles as a new tool for single biomolecule spectroscopy, *J Phys Chem B,* 105(48), 12165–12170.
15. Braun M, Cichos F (2013). Optically controlled thermophoretic trapping of single nano-objects, *ACS Nano,* 7(12), 11200–11208. http://pubs.acs.org/doi/10.1021/nn404980k.
16. Butler JE (2000). Solid supports in enzyme-linked immunosorbent assay and other solid-phase immunoassays, *Methods,* 22(1), 4–23. http://www.ncbi.nlm.nih.gov/pubmed/11020313%5Cnhttp://linkinghub.elsevier.com/retrieve/pii/S1046202300910314.
17. Cang H, et al. (2006). Confocal three dimensional tracking of a single nanoparticle with concurrent spectroscopic readouts, *Appl Phys Lett,* 88(22).
18. Cang H, Xu CS, Yang H (2008). Progress in single-molecule tracking spectroscopy, *Chem Phys Lett,* 457, 285–291.

19. Carlson CA, Sweeney NL, Nasse MJ, Woehl JC (2010). The corral trap: fabrication and software development, In *Proc SPIE 7571, Single Molecule Spectroscopy and Imaging III*, 757108.

20. Chandradoss SD, et al. (2014). Surface passivation for single-molecule protein studies, *J Visual Exp,* (86), 4–11. http://www.pubmedcentral.nih.gov/articlerender.fcgi?artid=4179479&tool=pmcentrez&rendertype=abstract.

21. Chen Y-F, et al. (2012). Controlled photonic manipulation of proteins and other nanomaterials, *Nano Lett,* 12(3), 1633–1637.

22. Cisse I, Okumus B, Joo C, Ha T (2007). Fueling protein DNA interactions inside porous nanocontainers, *Proc Natl Acad Sci USA,* 104(31), 12646–12650.

23. Cohen AE (2005). Control of nanoparticles with arbitrary two-dimensional force fields, *Phys Rev Lett,* 94(11), 118102.

24. Cohen AE (2006). *Trapping and Manipulating Single Molecules in Solution*, Stanford University.

25. Cohen AE, Moerner WE (2005). Method for trapping and manipulating nanoscale objects in solution, *Appl Phys Lett,* 86(9), 093109. http://aip.scitation.org/doi/10.1063/1.1872220.

26. Cohen AE, Moerner WE (2006). Suppressing Brownian motion of individual biomolecules in solution, *Proc Natl Acad Sci USA,* 103(12), 4362–4365.

27. Cohen AE, Moerner WE (2007). Principal-components analysis of shape fluctuations of single DNA molecules, *Proc Natl Acad Sci,* 104(31), 12622–12627. http://www.ncbi.nlm.nih.gov/pubmed/17496147 (January 13, 2017).

28. Cohen AE, Moerner WE (2008). Controlling Brownian motion of single protein molecules and single fluorophores in aqueous buffer, *Optics Express,* 16(10), 6941. https://www.osapublishing.org/oe/abstract.cfm?uri=oe-16-10-6941.

29. Cordes T, Vogelsang J, Tinnefeld T (2009). On the mechanism of trolox as antiblinking and antibleaching reagent, *J Am Chem Soc,* 131(14), 5018–5019. http://pubs.acs.org/doi/abs/10.1021/ja809117z.

30. Croce R, van Amerongen H (2014). Natural strategies for photosynthetic light harvesting, *Nat Chem Biol,* 10(7), 492–501.

31. Decher G (1997). Fuzzy nanoassemblies: toward layered polymeric multicomposites, *Science,* 277(5330), 1232–1237. http://www.sciencemag.org/cgi/doi/10.1126/science.277.5330.1232.

32. Deschout H, Neyts K, Braeckmans K (2012). The influence of movement on the localization precision of sub-resolution particles in fluorescence microscopy, *J Biophotonics,* 5(1), 97–109.

33. DeVore MS, et al. (2015). Note: time-gated 3D single quantum dot tracking with simultaneous spinning disk imaging, *Rev Sci Instrum,* 86(12), 126102. http://dx.doi.org/10.1063/1.4937477 (April 24, 2017).

34. Dienerowitz M, Dienerowitz F, Börsch M (2018). Measuring nanoparticle diffusion in an ABELtrap, *J Optics,* 20(3), 034006. https://doi.org/10.1088/2040-8986/aaa6fc (August 9, 2018).

35. Dienerowitz M, et al. (2016). Optimized green fluorescent protein fused to F o F 1 -ATP synthase for single-molecule FRET using a fast anti-Brownian electrokinetic trap, In J Enderlein et al. (eds), International Society for Optics and Photonics, 971402. http://proceedings.spiedigitallibrary.org/proceeding.aspx?doi=10.1117/12.2209592 (April 24, 2017).

36. Dovichi NJ, Zhang J (2000). How capillary electrophoresis sequenced the human genome, *Angew Chem Int Ed,* 39(24), 4463–4468. http://www.chem.ualberta.ca/~campbell/resources/Bioanalytical-2012/dovichi_review.pdf (September 4, 2017).

37. Dupont A, Lamb DC (2011). Nanoscale three-dimensional single particle tracking, *Nanoscale,* 3(11), 4532–4541.

38. Enderlein J (2000). Tracking of fluorescent molecules diffusing within membranes, *Appl Phys B,* 71(5), 773–777. http://link.springer.com/10.1007/s003400000409.

39. English BP, et al. (2006). Ever-fluctuating single enzyme molecules: Michaelis-Menten equation revisited, *Nat Chem Biol,* 2(2), 87–94. http://www.nature.com/doifinder/10.1038/nchembio759 (October 18, 2016).

40. Fields AP, Cohen AE (2011). Electrokinetic trapping at the one nanometer limit, *Proc Natl Acad Sci USA,* 108(22), 8937–8942. http://www.pubmedcentral.nih.gov/articlerender.fcgi?artid=3107292&tool=pmcentrez&rendertype=abstract.

41. Fields AP, Cohen AE (2012). Optimal tracking of a Brownian particle, *Opt Express,* 20(20), 22585–22601.

42. Foote AK, et al. (2019). Time-resolved multirotational dynamics of single solution-phase tau proteins reveals details of conformational variation, *Phys Chem Chem Phys,* 21(4), 1863–1871. https://pubs.rsc.org/en/content/articlehtml/2019/cp/c8cp06971a (August 24, 2020).

43. Fordyce PM, Valentine MT, Block SM (2008). Advances in surface-based assays for single molecules, In *Single-Molecule Techniques: A Laboratory Manual*, Cold Spring Harbor Laboratory Press.

44. Friedel M, Baumketner A, Shea JE (2006). Effects of surface tethering on protein folding mechanisms, *Proc Natl Acad Sci USA,* 103(22), 8396–8401.

45. Friedman LJ, Chung J, Gelles J (2006). Viewing dynamic assembly of molecular complexes by multi-wavelength single-molecule fluorescence, *Biophys J,* 91(3), 1023–1031. http://www.ncbi.nlm.nih.gov/pmc/articles/PMC1563747/.

46. Gavrilov, M, Bechhoefer J (2017). Feedback traps for virtual potentials, *Philos Trans R Soc London A, Math, Phys Eng Sci,* 375(2088). http://rsta.royalsocietypublishing.org/content/375/2088/20160217#ref-14 (May 4, 2017).

47. Germann JA, Davis LM (2014). Three-dimensional tracking of a single fluorescent nanoparticle using four-focus excitation in a confocal microscope, *Optics Express,* 22(5), 5641. https://www.osapublishing.org/oe/abstract.cfm?uri=oe-22-5-5641 (May 2, 2017).

48. Gitlin I, Carbeck JD, Whitesides GM (2006). Why are proteins charged? Networks of charge–charge interactions in proteins measured by charge ladders and capillary electrophoresis, *Angew Chem Int Edi,* 45(19), 3022–3060. http://doi.wiley.com/10.1002/anie.200502530.

49. Goldsmith RH, et al. (2011). Redox cycling and kinetic analysis of single molecules of solution-phase nitrite reductase, *Proc Natl Acad Sci USA,* 108(42), 17269–17274. http://www.pubmedcentral.nih.gov/articlerender.fcgi?artid=3198337&tool=pmcentrez&rendertype=abstract.

50. Goldsmith RH, Moerner WE (2010). Watching conformational- and photodynamics of single fluorescent proteins in solution, *Nat Chem,* 2(3), 179–186. http://www.nature.com/articles/nchem.545.

51. Goloborodko A, Marko JF, Mirny LA (2016). Chromosome compaction by active loop extrusion, *Biophys J,* 110(10), 2162–2168. http://linkinghub.elsevier.com/retrieve/pii/S0006349516301692 (May 4, 2017).

52. Ha T, et al. (1996). Probing the interaction between two single molecules: fluorescence resonance energy transfer between a single donor and a single acceptor, *Proc Natl Acad Sci U S A,* 93: 6264. http://www.pnas.org/content/93/13/6264.full.pdf (September 4, 2017).

53. Ha T, Kozlov AG, Lohman TM (2012). Single-molecule views of protein movement on single-stranded DNA, *Annu Rev Biophys,* 41: 295–319. http://www.pubmedcentral.nih.gov/articlerender.fcgi?artid=3719979&tool=pmcentrez&rendertype=abstract.

54. Ha T, Tinnefeld P (2012). Photophysics of fluorescent probes for single-molecule biophysics and super-resolution imaging, *Annu Rev Phys Chem,* 63(1), 595–617. http://www.annualreviews.org/doi/10.1146/annurev-physchem-032210-103340.

55. Hell SW, et al. (2015). The 2015 super-resolution microscopy roadmap, *J Phys D Appl Phys,* 48(44), 443001. http://stacks.iop.org/0022-3727/48/i=44/a=443001?key=crossref.64b10c5e92be3dbad6191dbfa3f07386 (October 17, 2016).

56. Hess ST, Girirajan TPK, Mason MD (2006). Ultra-high resolution imaging by fluorescence photoactivation localization microscopy, *Biophys J,* 91(11), 4258–4272.

57. Holmstrom ED, Nesbitt DJ (2010). Real-time infrared overtone laser control of temperature in picoliter H_2O samples: "nanobathtubs" for single molecule microscopy, *J Phys Chem Lett,* 1(15), 2264–2268.

58. Hoskins AA, et al. (2011). Ordered and dynamic assembly of single spliceosomes, *Science,* 331(6022), 1289–1295. http://www.sciencemag.org/cgi/doi/10.1126/science.1198830.

59. Hua B, et al. (2014). An improved surface passivation method for single-molecule studies, *Nat Methods,* 20(August), 1–7. http://www.ncbi.nlm.nih.gov/pubmed/25306544.

60. Hwang H, Myong S (2014). Protein induced fluorescence enhancement (PIFE) for probing protein–nucleic acid interactions, *Chem Soc Rev,* 43(4), 1221–1229. http://dx.doi.org/10.1039/C3CS60201J%5Cnhttp://xlink.rsc.org/?DOI=C3CS60201J.

61. Jiang Y, et al. (2011). Sensing cooperativity in ATP hydrolysis for single multisubunit enzymes in solution, *Proc Natl Acad Sci U S A,* 108(41), 16962–16967.

62. Joo C, et al. (2008). Advances in single-molecule fluorescence methods for molecular biology, *Annu Rev Biochem,* 77(1), 51–76. http://www.annualreviews.org/doi/10.1146/annurev.biochem.77.070606.101543 (October 17, 2016).

63. Juette MF, Bewersdorf J (2010). Three-dimensional tracking of single fluorescent particles with submillisecond temporal resolution, *Nano Lett,* 10(11), 4657–4563.

64. Juette MF, Rivera-Molina FE, Toomre DK, Bewersdorf J (2013). Adaptive optics enables three-dimensional single particle tracking at the sub-millisecond scale, *Appl Phys Lett,* 102(17), 173702. http://aip.scitation.org/doi/10.1063/1.4803538 (May 2, 2017).

65. Jun Y, Bechhoefer J (2012). Virtual potentials for feedback traps, *Phys Rev E,* 86(6), 061106. http://link.aps.org/doi/10.1103/PhysRevE.86.061106 (October 25, 2016).

66. Jun Y, Gavrilov M, Bechhoefer J (2014). High-precision test of Landauer's principle in a feedback trap, *Phys Rev Lett,* 113(19), 190601. https://link.aps.org/doi/10.1103/PhysRevLett.113.190601 (April 24, 2017).

67. Kartalov EP, Unger MA, Quake SR (2003). Polyelectrolyte surface interface for single-molecule fluorescence studies of DNA polymerase, *BioTechniques,* 34(3), 505–510.

68. Katayama Y, et al. (2009). Real-time nanomicroscopy via three-dimensional single-particle tracking, *ChemPhysChem,* 10(14), 2458–2464.

69. Kayci M, Chang HC, Radenovic A (2014). Electron spin resonance of nitrogen-vacancy defects embedded in single nanodiamonds in an ABEL trap, *Nano Lett,* 0(0): null.

70. Kayci M, Radenovic R (2015). Single florescent nanodiamond in a three dimensional ABEL trap, *Sci Rep,* 5(1), 16669. www.nature.com/scientificreports (August 14, 2019).

71. Keller AM, et al. (2012). Dynamic multibody protein interactions suggest versatile pathways for copper trafficking, *J Am Chem Soc,* 134(21), 8934–8943.

72. Keller AM, et al. (2014). 3-dimensional tracking of non-blinking 'giant' quantum dots in live cells, *Adv Funct Mater,* 24(30), 4796–4803.

73. King JK, Canfield BK, Davis LM (2013). Three-dimensional anti-Brownian electrokinetic trapping of a single nanoparticle in solution, *Appl Phys Lett,* 103(4), 43102. http://scitation.aip.org/content/aip/journal/apl/103/4/10.1063/1.4816325.

74. Kondo T, Chen WJ, Schlau-Cohen GS (2017). Single-molecule fluorescence spectroscopy of photosynthetic systems, *Chem Rev,* acs.chemrev.6b00195. http://pubs.acs.org/doi/abs/10.1021/acs.chemrev.6b00195 (January 10, 2017).

75. Krishnan M, Mojarad N, Kukura P, Sandoghdar V (2010). Geometry-induced electrostatic trapping of nanometric objects in a fluid, *Nature,* 467(7316), 692–695. http://dx.doi.org/10.1038/nature09404.

76. Kühn S, et al. (2009). Loss-based optical trap for on-chip particle analysis, *Lab Chip,* 9(15), 2212. http://pubs.rsc.org/en/content/articlepdf/2009/lc/b900555b (April 25, 2017).

77. Kuzyk A, et al. (2008). Dielectrophoretic trapping of DNA origami, *Small,* 4(4), 447–450.

78. Larson JD, Rodgers ML, Hoskins AA (2014). Visualizing cellular machines with colocalization single molecule microscopy, *Chem Soc Rev,* 43(4), 1189–1200. http://www.ncbi.nlm.nih.gov/pubmed/23970346.

79. Lesoine JF, et al. (2012). Nanochannel-based single molecule recycling, *Nano Lett,* 12(6), 3273–3278.

80. Lessard GA, Goodwin PM, Werner JH (2007). Three-dimensional tracking of individual quantum dots, *Appl Phys Lett,* 91(22), 224106. http://aip.scitation.org/doi/10.1063/1.2819074.

81. Limouse C, et al. (2017). Intramolecular dynamics of single molecules in free diffusion, *bioRxiv*, http://biorxiv.org/content/early/2017/03/24/120311 (May 2, 2017).

82. Liu Y, et al. (2013). Molecular orientation of enzymes attached to surfaces through defined chemical linkages at the solid-liquid interface, *J Am Chem Soc,* 135(34), 12660–12669.

83. Liu Z, Lavis LD, Betzig E (2015). Imaging live-cell dynamics and structure at the single-molecule level, *Mol Cell,* 58(4), 644. http://dx.doi.org/10.1016/j.molcel.2015.02.033.

84. Manger LH, et al. (2017). Revealing conformational variants of solution-phase intrinsically disordered tau protein at the single-molecule level, *Angew Chem Int Ed,* 56(49), 15584–15588. http://doi.wiley.com/10.1002/anie.201708242 (January 22, 2018).

85. McHale K, Mabuchi H (2009). Precise characterization of the conformation fluctuations of freely diffusing DNA: beyond rouse and zimm, *J Am Chem Soc,* 131(49), 17901–17907. http://pubs.acs.org/doi/abs/10.1021/ja906979j.

86. Michalet X (2010). Mean square displacement analysis of single-particle trajectories with localization error: Brownian motion in an isotropic medium, *Phys Rev E,* 82(4), 41914. https://journals.aps.org/pre/pdf/10.1103/PhysRevE.82.041914 (December 11, 2017).

87. Milanova D, Chambers RD, Bahga SS, Santiago JG (2012). Effect of PVP on the electroosmotic mobility of wet-etched glass microchannels, *Electrophoresis,* 33(21), 3259–3262.

88. Moerner WE (1999). Illuminating single molecules in condensed matter, *Science,* 283(5408), 1670–1676. https://www.sciencemag.org/lookup/doi/10.1126/science.283.5408.1670.

89. Moerner WE (2007). New directions in single-molecule imaging and analysis, *Proc Natl Acad Sci USA,* 104(31), 12596–12602.

90. Moerner WE, Shechtman Y, Wang Q (2015). Single-molecule spectroscopy and imaging over the decades, *Faraday Discuss,* 184(0), 9–36. http://pubs.rsc.org/en/Content/ArticleLanding/2015/FD/C5FD00149H (October 17, 2016).

91. Moffitt JR, Chemla YR, Smith SB, Bustamante C (2008). Recent advances in optical tweezers, *Annu Rev Biochem,* 77(1), 205–228. http://arjournals.annualreviews.org/doi/abs/10.1146/annurev.biochem.77.043007.090225?url_ver=Z39.88-2003&rfr_id=ori:rid:crossref.org&rfr_dat=cr_pub%3Dncbi.nlm.nih.gov.

92. Nie S, Chiu D, Zare R (1994). Probing individual molecules with confocal fluorescence microscopy, *Science,* 266(5187), 1018–1021. http://www.sciencemag.org/cgi/doi/10.1126/science.7973650.

93. Pang Y, Gordon R (2012). Optical trapping of a single protein, *Nano Lett,* 12(1), 402–406.

94. Perillo EP, et al. (2015). Deep and high-resolution three-dimensional tracking of single particles using nonlinear and multiplexed illumination, *Nat Commun,* 6(91), 7874. http://www.nature.com/ncomms/2015/150729/ncomms8874/full/ncomms8874.html.

95. Proesmans K, et al. (2016). Brownian duet: a novel tale of thermodynamic efficiency, *Phys Rev X,* 6(4), 041010. http://link.aps.org/doi/10.1103/PhysRevX.6.041010 (October 25, 2016).

96. Rahman M, Stott MA, Hawkins AR, Schmidt H (2016). Design and characterization of integrated 2D ABEL trap, In *2016 IEEE Photonics Conference (IPC)*, IEEE, 374–375. http://ieeexplore.ieee.org/document/7831143/ (April 25, 2017).

97. Rahmanseresht S, et al. (2015). Single-molecule-sensitive fluorescence resonance energy transfer in freely-diffusing attoliter droplets, *Appl Phys Lett,* 106(19), 194107. http://scitation.aip.org/content/aip/journal/apl/106/19/10.1063/1.4921202 (October 18, 2016).

98. Rasnik I, McKinney SA, Ha T (2005). Surfaces and orientations: much to FRET about?, *Acc Chem Res,* 38(7), 542–548.

99. Rasnik I, McKinney SA, Ha T (2006). Nonblinking and longlasting single-molecule fluorescence imaging, *Nat Methods,* 3(11), 891–893.

100. Ries J, Schwille P (2012). Fluorescence correlation spectroscopy, *BioEssays,* 34(5), 361–368. http://doi.wiley.com/10.1002/bies.201100111 (October 17, 2016).

101. Rigler R, Elson ES (2012). *Fluorescence Correlation Spectroscopy: Theory and Applications*, Springer, Berlin Heidelberg, https://books.google.com/books?id=iAH8CAAAQBAJ.

102. Rissin DM, Walt DR (2006). Digital concentration readout of single enzyme molecules using femtoliter arrays and poisson statistics, *Nano Lett,* 6(3), 520–523.

103. Rondelez Y, et al. (2005). Microfabricated arrays of femtoliter chambers allow single molecule enzymology, *Nat Biotechnol,* 23(3), 361–365.

104. Roy R, Hohng S, Ha T (2008). A practical guide to single-molecule FRET, *Nat Methods,* 5(6), 507–516.

105. Rubinovich L, Polak, M (2013). The intrinsic role of nanoconfinement in chemical equilibrium: evidence from DNA hybridization, *Nano Lett,* 13(5), 2247–2251.

106. Rust MJ, Bates M, Zhuang X (2006). Sub-diffraction-limit imaging by stochastic optical reconstruction microscopy (STORM), *Nat Methods,* 3(10), 793–796.

107. Sahl SJ, Moerner WE (2013). Super-resolution fluorescence imaging with single molecules, *Curr Opin Struct Biol,* 23(5), 778–787. http://linkinghub.elsevier.com/retrieve/pii/S0959440X13001437 (October 17, 2016).

108. Schlau-Cohen GS, et al. (2013). Single-molecule spectroscopy reveals photosynthetic LH2 complexes switch between emissive states, *Proc Natl Acad Sci,* 110(27), 10899–10903. http://www.pnas.org/content/110/27/10899.

109. Schlau-Cohen GS, et al. (2015). Single-molecule identification of quenched and unquenched states of LHCII, *J Phys Chem Lett,* 6(5), 860–867. http://pubs.acs.org/doi/abs/10.1021/acs.jpclett.5b00034 (January 10, 2017).

110. Schlau-Cohen GS, Bockenhauer S, Wang Q, Moerner WE (2014). Single-molecule spectroscopy of photosynthetic proteins in solution: exploration of structure–function relationships, *Chem Sci,* 5(8), 2933–2939. http://pubs.rsc.org/en/content/articlehtml/2014/sc/c4sc00582a.

111. Schuler B (2013). Single-molecule FRET of protein structure and dynamics—a primer, *J Nanobiotechnol,* 11(Suppl 1), S2. http://jnanobiotechnology.biomedcentral.com/articles/10.1186/1477-3155-11-S1-S2 (October 17, 2016).

112. Senavirathne G, et al. (2015). Widespread nuclease contamination in commonly used oxygen-scavenging systems, *Nat Methods,* 12(10), 901–902. http://www.nature.com/doifinder/10.1038/nmeth.3588.

113. Shapiro B, Chaudhary SV, Armani MD, Probst R (2005). Arbitrary and simultaneous control of multiple objects in microfluidic systems.

114. Shon MJ, Cohen AE (2012). Mass action at the single-molecule level, *J Am Chem Soc,* 134(35), 14618–14623. http://pubs.acs.org/doi/10.1021/ja3062425.

115. Squires AH, et al. (2019). Single-molecule trapping and spectroscopy reveals photophysical heterogeneity of phycobilisomes quenched by orange carotenoid protein, *Nat Commun,* 10(1), 1172. https://doi.org/10.1038/s41467-019-09084-2 (March 19, 2019).

116. Squires AH, Moerner WE (2017). Direct single-molecule measurements of phycocyanobilin photophysics in monomeric C-phycocyanin, *Proc Natl Acad Sci,* 114(37), 9779–9784. http://www.ncbi.nlm.nih.gov/pubmed/28847963 (September 4, 2017).

117. Su B, et al. (2015). Observing conformations of single F o F 1 -ATP synthases in a fast anti-Brownian electrokinetic trap, *Proc SPIE,* 9329, 93290A. http://proceedings.spiedigitallibrary.org/proceeding.aspx?doi=10.1117/12.2080975.

118. Sui G, et al. (2006). Solution-phase surface modification in intact poly(dimethylsiloxane) microfluidic channels, *Anal Chem,* 78(15), 5543–5551.

119. Sustarsic M, Kapanidis AN (2015). Taking the ruler to the jungle: single-molecule FRET for understanding biomolecular structure and dynamics in live cells, *Curr Opin Struct Biol,* 34, 52–59. http://linkinghub.elsevier.com/retrieve/pii/S0959440X15000913 (October 17, 2016).

120. Swoboda M, et al. (2012). Enzymatic oxygen scavenging for photostability without Ph drop in single-molecule experiments, *ACS Nano,* 6(7), 6364–6369.

121. Vogelsang J, et al. (2008). A reducing and oxidizing system minimizes photobleaching and blinking of fluorescent dyes, *Angew Chem Int Ed,* 47(29), 5465–5469. http://doi.wiley.com/10.1002/anie.200801518.

122. Wang Q, et al. (2012). Probing single biomolecules in solution using the Anti-Brownian ELectrokinetic (ABEL) trap, *Acc Chem Res,* 45(11), 1955–1964. http://pubs.acs.org/doi/abs/10.1021/ar200304t.

123. Wang Q, et al. (2014). Enabling multivariate investiation of single-molecule dynamics in solution by counteracting brownian motion, Stanford University.

124. Wang Q, Moerner WE (2010). Optimal strategy for trapping single fluorescent molecules in solution using the ABEL trap, *Appl Phys, B* 99(1–2), 23–30. http://link.springer.com/10.1007/s00340-009-3843-y.

125. Wang Q, Moerner WE (2011). An adaptive anti-Brownian electrokinetic trap with real-time information on single-molecule diffusivity and mobility, *ACS Nano,* 5(7), 5792–5799. http://pubs.acs.org/doi/abs/10.1021/nn2014968.

126. Wang Q, Moerner WE (2013). Lifetime and spectrally resolved characterization of the photodynamics of single fluorophores in solution using the anti-Brownian electrokinetic trap, *J Phys Chem B,* 117(16), 4641–4648. http://pubs.acs.org/doi/abs/10.1021/jp308949d.

127. Wang Q, Moerner WE (2014a). Single-molecule motions enable direct visualization of biomolecular interactions in solution, *Nat Methods,* 11(5): 555–558. http://www.nature.com/doifinder/10.1038/nmeth.2882.

128. Wang Q, Moerner WE (2014b). Spectroscopic and transport measurements of single molecules in solution using an electrokinetic trap, In *Proc SPIE,* J Enderlein et al. (eds), 895004. http://proceedings.spiedigitallibrary.org/proceeding.aspx?doi=10.1117/12.2038320.

129. Wang Q, Moerner WE (2015). Dissecting pigment architecture of individual photosynthetic antenna complexes in solution, *Proc Natl Acad Sci,* 112(45), 13880–13885. http://www.pnas.org/lookup/doi/10.1073/pnas.1514027112.

130. Watkins LP, Yang H (2005). Detection of intensity change points in time-resolved single-molecule measurements, *J Phys Chem B,* 109(1), 617–628.

131. Welch G, Bishop G (1995). An Introduction to the Kalman Filter.

132. Welsher K, Yang H (2014). Multi-resolution 3D visualization of the early stages of cellular uptake of peptide-coated nanoparticles, *Nat Nanotechnol,* 9(3), 198–203. http://www.nature.com/doifinder/10.1038/nnano.2014.12.

133. Widengren J, et al. (2006). Single-molecule detection and identification of multiple species by multiparameter fluorescence detection, *Anal Chem,* 78(6), 2039–2050. http://pubs.acs.org/doi/abs/10.1021/ac0522759.

134. Wunderlich B, et al. (2013). Microfluidic mixer designed for performing single-molecule kinetics with confocal detection on timescales from milliseconds to minutes, *Nat Protoc,* 8(8), 1459–1474. http://www.nature.com/doifinder/10.1038/nprot.2013.082 (July 20, 2016).

135. Yang AHJ, et al. (2009). Optical manipulation of nanoparticles and biomolecules in sub-wavelength slot waveguides, *Nature,* 457(7225), 71–75. http://www.nature.com/doifinder/10.1038/nature07593.

136. Yang H, et al. (2003). Protein conformational dynamics probed by single-molecule electron transfer, *Science,* 302(5643), 262–266.

137. Yildiz A (2003). Myosin V walks hand-over-hand: single fluorophore imaging with 1.5-Nm localization, *Science,* 300(5628), 2061–2065. https://www.sciencemag.org/lookup/doi/10.1126/science.1084398.

138. Ying L, Xie XS (1998). Fluorescence spectroscopy, exciton dynamics, and photochemistry of single allophycocyanin trimers, *J Phys Chem B,* 102(50), 10399–10409.

139. Zhao Z, et al. (2016). Nanocaged enzymes with enhanced catalytic activity and increased stability against protease digestion, *Nat Commun,* 7, 10619. http://www.nature.com/doifinder/10.1038/ncomms10619 (October 24, 2016).

140. Zheng Q, et al. (2014). Ultra-stable organic fluorophores for single-molecule research, *Chem Soc Rev,* 43(4), 1044–1056. http://xlink.rsc.org/?DOI=C3CS60237K (April 28, 2017).

Index